ASPECTS OF VAGUENESS

THEORY AND DECISION LIBRARY

AN INTERNATIONAL SERIES
IN THE PHILOSOPHY AND METHODOLOGY OF THE
SOCIAL AND BEHAVIORAL SCIENCES

VOLUME 39

ASPECTS OF VAGUENESS

Edited by

HEINZ J. SKALA
University of Paderborn, F.R.G.

S. TERMINI
Consiglio Nazionale delle Ricerche, Laboratorio di Cibernetica, Naples, Italy

and

E. TRILLAS
Universidad Politécnica de Barcelona, Spain

D. REIDEL PUBLISHING COMPANY

A MEMBER OF THE KLUWER ACADEMIC PUBLISHERS GROUP

DORDRECHT / BOSTON / LANCASTER

Library of Congress Cataloging in Publication Data

CIP

Aspects of vagueness

(Theory and Decision Library = v. 39)

Papers presented at the Second World Conference on Mathematics at the Services of Man, held at the Universidad Politecnica de Las Palmas, Canary Islands, Spain, June 23 to July 3, 1982, and sponsored by the Cabildo Insular de Gran Canaria and the Universidad.

Includes indexes.

1. Fuzzy sets-Congresses. 2. Fuzzy Systems-Congresses. 3. Mathematical Models-Congresses. I. Skala, Heinz J. II. Fermini, S. (Settimo), 1945- III. Trillas, E. (Enric) IV. Conference on mathematics at the services of man (2nd: 1982: Universidad Politecnica de Las Palmas) V. Gran Canaria (Canary Islands). Cabildo Insular, VI. Universidad Politecnica de Las Palmas. VII. Title: Vagueness. VIII. Series.

QA248.A87 1984 508.3'2 83-26994

ISBN 90-277-1692-7

Published by D. Reidel Publishing Company,
P.O. Box 17, 3300 AA Dordrecht, Holland.

Sold and distributed in the U.S.A. and Canada
by Kluwer Academic Publishers,
190 Old Derby Street, Hingham, MA 02043, U.S.A.

In all other countries, sold and distributed
by Kluwer Academic Publishers Group,
P.O. Box 322, 3300 AH Dordrecht, Holland.

Printed in The Netherlands

TABLE OF CONTENTS

PREFACE

The Second World Conference on Mathematics at the Service of Man was held at the Universidad Politécnica de Las Palmas, Canary Islands, Spain, June 28 to July 3, 1982. The first volume of the Proceedings of the Conference, entitled "Functional Equations-Theory and Applications" has appeared in the Reidel series "Mathematics and Its Applications". The papers in this volume consist of the invited lectures delivered at the Conference, Section 7: Non-Classical Logics and Modelling, as well as some selected papers which offer an introduction to the philosophy, methodology and to the literature of the broad and fascinating field of vagueness, imprecision and uncertainty. The contributed papers appeared in the volume of photo-offset preprints distributed at the Conference.

It is our hope that the papers present a good sample with respect to the background, the formalism and practice of this area of research as far as we understand it today. As the subject "Vagueness" touches many aspects of human thinking, the contributions have been made from a broad spectrum ranging from philosophy through pure mathematics to probability theory and mathematical economics, therefore the careful reader should find some new insights here.

In conclusion, the editors want to thank all authors who have contributed to this volume; the publishers of "Commentationes Mathematicae Universitatis Carolinae" for permission to reprint the paper "Fuzziness and Fuzzy Equality", Commentationes Mathematicae Universitatis Carolinae 23 (1982), 249-267, and D. Reidel for friendly cooperation.

We would also like to thank the sponsors of the conference: the Cabildo Insular de Gran Canaria, and the Universidad Politécnica de Las Palmas. One of the editors was supported by the Deutsche Forschungsgemeinschaft. Last, but not least, we wish to express our appreciation to Mrs. J. Höxtermann, for secretarial assistance and K. Braun, Paderborn.

The Editors

H. J. Skala, S. Termini, and E. Trillas (eds.), Aspects of Vagueness, vii.

Jiří Bečvář

NOTES ON VAGUENESS AND MATHEMATICS

1. This paper consists of a list of notes on the issue of vagueness, the emphasis being on its relation to "crisp" entities. This is a classical topic, in fact any examination of the interplay of empirical material and theoretical concepts has to take it into account (see,e.g., [4]). In addition to earlier investigations, last decades have brought substantial contributions in the direction of providing technical mathematical and logical explications of aspects of vagueness. This development has been in part initiated by problems in the theory of systems, in computer science, in the area of artificial intelligence, in foundations of mathematics, etc. The most extensive branch of the development has been the theory of fuzzy sets, originating with [9] (for a detailed overview, see [2] and the bibliography in [3]). Among important contributions based on different principles let us mention especially the alternative set theory [8].

The exposition in this paper does not lean upon a concrete technical explication of the concept of vagueness. Also the language used is deliberately (mostly) nontechnical; thus, in particular, we use the word "vague" rather as a generic term, encompassing various shades of being imprecise, indeterminate, fuzzy, ambiguous, uncertain, etc. This corresponds to the orientation of the paper: its aim is to recall selected aspects of vagueness, sometimes in less frequently discussed connections which, nevertheless, may prove relevant to the formation of a general background. (The whole picture is composed of essentially familiar items; accordingly, only a small number of bibliographical references have been included.)

2. There is a striking contrast between our ablility in everyday practice to successfully recognize objects, events, properties etc., and, on the other hand, our inability to provide complete descriptions or unambiguous characterizations of them. Experience tells us that any attempt at a complete characterization of a piece of reality eventually reaches a point of uncertainty as to the adequacy and/or applicability of the identification criteria suggested. In

H. J. Skala, S. Termini, and E. Trillas (eds.), Aspects of Vagueness, 1–11.

what follows we will look more closely at some of the ingredients of this bipolar situation.

The world in which we live is not homogeneous; it exhibits, on all levels of magnitude and development, some relatively stable patterns. The articulation is especially appealing in the organic sphere. On the highest level known to us so far, human organisms have gradually developed, in a continuous interaction with the environment, the ability to isolate, both in the external world and in the sphere of conscious thinking, a considerable number of more or less definite entities, pertaining to all branches of human activity (objects, relations, states, propositions, processes, concepts of utility, causality, mathematical concepts ...; properties of being complex, abstract, vague also belong here).

In these entities we are able to recognize the presence of various degrees of vagueness. At the same time we realize that some of them, especially the basic logical and mathematical constructs, appear to us as little vague as is conceivable at all [7], at least in the moment when they become involved in the central zone of our mental activity.

In general, the status of being vague is influenced by the context in which an entity appears. For example, the process and result of an interaction between vaguely determined lumps of, say, organic matter can be sometimes adequately thought of in a rather crisp way. Also the human routine identification and manipulation of objects proceeds so that we seem to experience them directly as sufficiently clear and have the impression of handling them according to simple rules.

3. Our brain itself accomodates an ingredient which provides an active contraposition to vagueness. It consists in that entities once conceived in it can be treated there also in a formal, or "combinatorial", way, largely independent of their source, original meaning and imprecision; they - or the words used for them - can even play the role of mere auxiliary carriers of modified hypothetical meanings which may be incomplete, in different senses vague, or contradictory (similar situation obtains or can be simulated in a computer) and may this reach the level and status which is typical of operating with abstract objects within the framework of mathematics and (classical) logic. We realize that

also in the current use and understanding of a natural language - which, with the probable exception of its "logical" kernel, typically reflects the complicated, non-crisp part of our mental processes - elements of the combinatorial attitude are operative. (For example, when listening to a long phrase one uses to solve the problem of understanding the whole by making, among others, during hearing a part of the phrase a quick mental experimentation and predictions as to what the next parts may be, in order that the task is completed without a disturbing delay. Moreover, during listening only some "representative" features of the meanings of the words and phrases are being called up, as possible stimuli for a subsequent analysis, inquiry into the truth, consistency, etc.)

4. We mentioned that the mathematical and logical frame seems to us to have supreme clarity and precision. This does not imply, however, that its basic elements are simple or "prime". As a matter of fact, they appear to be located in the uppermost levels of an immense pyramid of complex automatic and selforganizing processes, the structure, evolution and interpretation of which are still largely unknown. (For a parallel let us recall that our conscious interpretation of what appears in the visual field is also preceded by a vast, unconscious analysis and preprocessing of the patterns on the retina.) It would not be much enlighting thus to say that the mathematical frame is vague. We prefer just to conclude that basing an explication of the phenomena in the real world on a reduction to this frame is possibly a sort of high-level translation. Accordingly, both the questions about its directness and universal explicative capacity are open. This proves relevant, in particular, to the attempts at an explication of various features of vagueness, e.g. the dichotomy discrete-continuous, etc.; an "adequate" representation of phenomena in the sphere of elementary physical particles is a related problem.

(For a computer, the underlying (technological) pyramid is different. Hence also the question about the explicative/ modelling capacity of contemporary computers - which are sophisticated artificial arrangements of pieces of matter, subject to physical laws which are sufficiently well known for the intended purpose - points in a somewhat different direction. Let us note, incidentally, that computers are not Garden-of-Eden configurations in the complete system of Nature; they appear within it as the result of its own motion.)

There are still other important elements which make the status of mathematical entities relative. One of them is of a subjective, or, more properly speaking, psychological, nature: A condition sine qua non for being able to operate efficiently with abstract mathematical entities is to experience ("see") them clearly - clare et distincte -, nearly as concrete objects. The corresponding state of mind cannot be communicated or completely described. (In this respect it is similar to a large number of phenomena like the fusion of two slightly different images in binocular vision, sensing the harmonic quality of a chord as different from the notes constituting it, etc.)

Another, similar element, but on a somewhat higher level, is manifest if we try to use mathematics and logic as tools for the explication of new phenomena; in particular, for an analysis of some basic concepts of mathematics themselves. This proves impossible unless we sufficiently clearly "understand" what we are doing. Such an understanding, especially if connected with the establishment of a new conceptual base, is usually not the result of a merely deductive or combinatorial manipulation of old concepts, but is simultaneously related to new insights and intuitions - and decisions; these then result in a collective understanding, specific for a group of people or an epoch. (As a rule, in this process some old, delicate interpretative questions are left unresolved, but the attraction and success of the new approach prevail. Generations of students have learnt and assimilated mathematical analysis in arithmetical form, set theory in Cantor's spirit, probability theory in Kolmogorov's axiomatization ...)

5. In the articulated real world many of the entities interacting on a certain level are aggregates of a very large number of elements of lower levels; we may loosely call them macroentities or macroobjects (of that level). For the result of the interaction, variations of their detailed structure are, within certain limits, without substantial effect. (In the organic sphere this is stressed by measures taken to count with the possibility of errors and malfunction of elements or parts.) Our life is adapted to the presence of vagueness; it both respects the resulting restrictions and

profits from them. For example, the question of a precise identification of a macroentity (person, coin, storm) is practically never posed, even if we are aware of its many details or parts; we have the impression of recognizing it by a few "representative" features and on the basis of some global characteristics (which, however, can be arrived at by a rather complicated subconscious activity of the organs of our body).

In some bottleneck cases, a catastrophe may happen if our behaviour is not rather precisely appropriate for the situation, but there are presumably not many problems (if any) which have to be solved with large precision during a "practically" unlimited period of time and over a "practically" unlimited region in space by a single object/machine or organism. As a rule, for solving a difficult or new problem, enough time is available in advance to complete the necessary training, testing of new methods, fabrication of sufficiently precise and powerful instruments, etc. (The situation here is similar to that which obtains in regard to the limitations and problems imposed by the phenomena of complexity; cf. [1].)

6. We are aware that for a large majority of questions which we are able to formulate (especially for those formulated in natural language) we do not have an unambiguous answer. There are numerous reasons: our ignorance, we simply do not know; incompleteness of the meanings of the words, which do depend on the real situation in which the question is being put; intentional use of words with fuzzy meaning, having the purpose to indicate either a kind of impression or just to suggest a general orientation; etc. Sometimes the primary vagueness of a question (or of a predicate involved) is largely reduced if the question is followed by the contra-question Why do you ask? (As in the follwing dialogue. A: Is John tall? B: Why? A: I would like to know if this skis are long enough for him. B: Oh, he is tall!) - Vague questions are being currently answered, usually also vaguely; modalities with highly vague meanings are a frequent ingredient. In many cases, we do not try to make a question more precise or to look for an answer; we just wait until this is somehow solved, if ever. If, however, there exists a sufficiently strong motivation, we initiate inquiry, experiments or scientific research for reaching a sufficient answer.

7. It is characteristic and important that, in order to devise and construct a machine or a system (hammer, computer), we need not know the motion of every atom being an element of it. It suffices to have the knowledge of global conditions (relating in general to several levels and being sometimes rather complex) which can call forth processes that will end with the existence and/or running of the system. These global conditions are frequently expressed in mathematical terms. In spite of the contents of Section 4, mathematical constructs retain for us the highest crispness; thus there is the problem in which relation they can stand to those global conditions, which are full of fuzzy details.

Let us look at the status of mathematical entities more closely. Mathematics is embedded via our heads into the real world. In the brain, together with the mathematical constructs that we experience so crisply, there float specimens of all sorts of our mental activity, including all types of reflections of the external world: extremely clear direct or imaginative pictures of real situations (the latter sometimes so crisp that we are able to isolate in them every tiny detail, like on a minute photograph); general global ideas of macroentities; words as potential generators of their meanings; (our experiences of) abstract concepts ...

Within mathematics, we can recognize the following important components:
(i) Finite combinatorics of (not very large) aggregates of crisp elements.
(ii) Idea of an iterative continuation (or generation) of a series of steps which are conceived as uniform.
(iii) Principle of (abstract) aggregation of somehow given entities (especially those having a given property).

A strengthening of point (ii) in a spirit similar to (iii) is responsible for the acceptance, in mathematics, of ideal limit elements. Point (iii) has direct connection to the concept of a set or class; it is also partly connected with our ability to adopt a "metaview" in regard to many entities or situations. (In a computer, this can take on very concrete forms.) The combination of (ii) and (iii) leads to the conception of the (classical) mathematical continuum.

In addition, there plays in mathematics (but not only there, cf. Section 4) a substantial role the hypothetical attidude:

(iv) Possibility of adopting an axiomatic point of view. This point provides the main tool for handling situations in the real world in which we have to do with objects, of which we do not know some properties, internal structure or details (e.g. a wire leading electric current), but suspect their existence and responsability for some (macro)phenomena which we can observe.

The mathematical framework (in the narrower sense) does not itself provide criteria for deciding whether a specific mathematical structure is appropriate for the description of a phenomenon in the real (in a special case: external) world. However, all information about the phenomenon is present in the brain, in contact with the mathematical structure, thus a confrontation in a common medium is possible. (For simplicity of the argument, we disregard the important possibility that the information is (partly) stored in a computer.) If the confrontation is not resultative, either more information may be required or a decision be made as to the acceptance of the proposed mathematical model of the phenomenon. It is decision, as well, which solved practically the problem of the possibility of an unlimited regress when tests are required for the adequacy of a set of axioms as a description of a phenomenon, then test of second order for the problem of adequacy of the tests of first order, etc.

In the case of a phenomenon (a certain aspect of vagueness may be an example), where neither the whole of our conceptual base nor the external information and experiments provide a sufficiently determinate answer to the question of adequacy of a proposed (axiomatic) explication; the act of decision may be essential, providing a new background and leading possibly to a new understanding and intuition. (We know, of course, that in the very practical life making just a decision may sometimes be at least as important as possessing a knowledge.)

8. A considerable number of structures and activities are involved (the hypothetical attitude and decisions playing a prominent role among them), when a mathematical description of a real situation is looked for - even if the situation is a static one, not involving motion of its microelements.

As an example, consider a photographic plate A, on which the distribution of a very large number of black silver grains is such that the global pattern which we see is, say, a sequence of parallel, equidistant lines. Let us assume that we are not able (without using, e.g., a microscope) to see clearly (or at all) the individual grains. Then we are interested especially in such mathematical descriptions, from which we can derive global features of the situation (for example the number of the lines, or the contrast of the picture at the edge of a line). In the old classical physics, one would probably not have hesitated to accept as basic that theoretical description in which each grain is localized by identifying its center of mass with a point in a planar mathematical continuum,etc. One would then jump from separate grains to a continuous density function, which in turn would have to be extrapolated from the results of measuring density on a finite number of sample places by means of instruments, capable to provide a global response without sensing the individual grains. In this treatment, a complete description of the distribution of the grains seems impossible since we used elements of the mathematical continuum (real numbers).

We find, however, that the impossibility obtains also if we do not explicitly introduce the structure of the continuum and stop on a level of detail which approximately corresponds to the size of the grains. We at once observe that the plate A cannot simply be identified with the set of grains; in the mathematical model, we need a substitute for the space in which the grains are located. A finite set G can serve the purpose, playing the role of a planar grid, so fine that with small error each grain can be represented by a mesh of G. The global picture which we actually see is also distributed in the plane, as a pattern composed of grey patches, the shade of grey being rather uniform within one patch. Let us again use a grid, G′, for the approximate localization of the patches, the size of the meshes of G′ and of the patches being approximately the same. To simplify the argument, let us assume that the grid G on the microlevel is a refinement of G′. According to our assumption as to the poor possibility to identify the grains, a mesh in G′ includes a very large number of meshes of G; its shade is determined by the number of grains placed in it via G. It is natural to assume that the distribution of the grains is not locally regular; thus the same macroeffect may be caused by an immense number of microstructures. We are able, when using a microscope, to differentiate and describe the inner distribution of grains in a portion of G which corresponds to a tiny piece of the global picture. We are, however, unable, for complexity reasons, to produce the record of the whole microstructure of the plate. Thus the real distributions of the grains form rather a family of theoretical possibilities how to explicate (explain) the picture seen. (Taking an example with a larger span between the micro- and macrolevels would make the effect more conspicuous. The use of computer memory and representation may, of course, shift the phenomenon several orders down or up.)

In our previous passage from one level to another we concentrated on the complexity problem. Let us mention also another, rather conceptual problem connected with vagueness that emerges when we look at a set A from a larger distance, so that the (micro)elements of A cease to be clearly distinguishable. (Russell[7] expressed the view that all phenomena of vagueness in language and thought are essentially analogous to this case; a similar standpoint is adopted in the alternative set theory [8] concerning the mathematical explication of continuum phenomena (cf. Remark below)). Mathematically, one uses to assume that a relation T of indistinguishability on A is present which is a tolerance, i.e. T is reflexive and symmetric (but no transitive). From the distant view, however, the microelements are no more legitimate, identifiable entities, thus to express the new situation we would need to replace them by some new mathematical entities. If we retain the original elements of A as ideal "atoms", then, classically, the new elements are chosen to be subsets of A. Since,however, T is not an equivalence relation, we have not corresponding equivalence classes available, thus not a natural hint for choosing new, larger elements than before. (The cliques of the relation T are generally not good candidates for them.)

(Mixing the points of view corresponding to different observational levels is a frequent source of paradoxes; for a discussion of them, see e.g. [5].)

Remark. The nontransitivity of the relation of indistinguishability was discussed already by Poincaré [6], as a characteristic feature of the physical continuum in contrast to the mathematical continuum.

In the alternative set theory (AST) one considers a countable (in the sense of AST) sequence of set-theoretically definable classes T_n, which are tolerances. Their intersection $\doteq$ is a class which is not only a tolerance, but is even transitive, thus an equivalence relation. If $\doteq$ has the additional property that each infinite (in the sense of AST) set contains different elements x,y such that $x \doteq y$, then $\doteq$ is called an indiscernibility equivalence. Such a $\doteq$ determines the fundamental microstructure for the study of continuum (and vagueness) phenomena in the AST.

The transitivity of $\doteq$ follows from the fact that in AST one can prove (using the so-called prolongation axiom) that the tolerances T_n can be assumed, after a modification, to satisfy for each n the condition $T_{n+1} \circ T_{n+1} \subseteq T_n$, where "∘" means the composition of relations. This is the condition of uniformity known from the theory of uniform spaces. There, however, it has rather to be postulated (in some analogous form; for tolerances which form basis of the space).

Mathematical Institute, Prague,
Czechoslovak Academy of Sciences

REFERENCES

[1] Bečvář, J.: 1965, 'Real-time and complexity problems in automata theory'. Kybernetica 1, 475-498.

[2] Gaines, B.R.: 1976, 'Foundations of fuzzy reasoning'. Int.J.Man-Machine Studies 8, 623-668.

[3] Gaines, B.R. and L.J.Kohout: 1977, 'The fuzzy decade: a bibliography of fuzzy systems and closely related topics'. Int.J.Man-Machine Studies 9, 1-68.

[4] Körner, S.: 1966, 'Experience and theory'. Routledge and Kegan Paul, London 1966.

[5] Parikh, R.: 1977, 'The problem of vague predicates'. Preprint.

[6] Poincaré, H.: 1912, 'La science et l'hypothèse'. Paris.

[7] Russel, B.: 1923, 'Vagueness'. Austr.J.of Phil. I, 84-92.

[8] Vopenca, P.: 1979, 'Mathematics in the alternive set theory'. Teubner, Leipzig.

[9] Zadeh, L.A.: 1965, 'Fuzzy sets'. Information and Control 8, 338-353.

Siegfried Gottwald

FUZZY SET THEORY: SOME ASPECTS OF THE EARLY DEVELOPMENT

1. INTRODUCTION

In a well-known study on connections of fuzzy set theory with many-valued logic, R. Giles (1976) refers to "a neglected series of papers by Klaua (1965-1970)", noticing that "Klaua ... develops a 'many-valued set theory' based on Łukasiewicz logic in a manner similar to (but much more sophisticated than) that adopted here. ... His work goes far beyond the present paper in the mathematical development of the subject and deserves much more attention than it appears to have received."

It seems that this situation has not changed much to date. Therefore, this paper aims at surveying the results reported in those papers and some development arising from them and realized mainly at the Karl Marx University, Leipzig, where Klaua held a full professorship from 1965 to 1972 and worked with a small group of young people, including the present author.

For a unified presentation we will use, besides the well-known notation of Zadeh for generalized membership values of fuzzy sets, some connectives and predicates of language of many-valued logic and as set $\underline{w}$ of truth values of the many-valued logic we will consider always the set of generalized membership values, i.e. the whole real interval $[0,1]$ or some finite subset of it one of the sets $w_n = \{k/(n-1) \mid 0 \leq k < n\}$. Let us write $\neg$, $\wedge$, $\vee$ as is usual for the many-valued connectives of negation, conjunction, disjunction defined by 1-.., minimum, maximum, respectively we use $\rightarrow$, $\leftrightarrow$ for Łukasiewicz implication and biimplication, and & for bold conjunction as in Giles(1976). Finally, $\forall$ and $\exists$ are used for many-valued generalization and existential quantification, given respectively by inf sup.
With $\langle P \rangle$ for the truth value of an expression P of our language of many-valued logic we thus have: $\langle \neg P \rangle = 1 - \langle P \rangle$, $\langle P \wedge Q \rangle = \min(\langle P \rangle, \langle Q \rangle)$, $\langle P \vee Q \rangle = \max(\langle P \rangle, \langle Q \rangle)$, $\langle P \& Q \rangle = \max(0, \langle P \rangle + \langle Q \rangle - 1)$, $\langle P \rightarrow Q \rangle = \min(1, 1 - \langle P \rangle + \langle Q \rangle)$ and $\langle P \leftrightarrow Q \rangle = 1 - |\langle P \rangle - \langle Q \rangle|$. Quantification is described by $\langle \forall x P(x) \rangle = \inf_a \langle P(a) \rangle$ and $\langle \exists x P(x) \rangle = \sup_a \langle P(a) \rangle$ for a suitable domain of

H. J. Skala, S. Termini, and E. Trillas (eds.), Aspects of Vagueness, 13–29.

quantification. Restricted quantification is used with crisp sets as restrictions only and written $\forall_M x$, $\exists_M x$ with interpretation by $\inf_{x \in M}$ resp. $\sup_{x \in M}$.

Additionally, we need the membership relation $\in$ of classical set theory, the class $\underline{On}$ of ordinals, the difference $M \setminus N$ of crisp sets M,N, the notions dom f, rg f of domain and range of a mapping f. We use the notation $(M_i \mid i \in I)$ for the family of all M_i with index from set I; and $\models$ indicates that the following formula of many-valued logic holds true, i.e. has (always) truth value 1.

2. TWO HIERARCHIES OF GENERALIZED SETS

2.1. Only membership and identity generalized

As explained in his works (1965, 1967a) and in direct conversation, D. Klaua got the impulse for his investigations from conversations with Karl Schröter and Karl Menger.
The last named, in particular, advised him that perhaps he could use for some of his studies a generalized membership relation admitting intermediate values between "true" and "false", i.e. between 1 and 0. From the very beginning it was the idea to also use the means of language of many-valued logic for describing such a generalization, and to consider for this purpose the generalized membership values as the generalized truth values.

Hence, the aim of these studies was very different of that of, e.g. Skolem (1957, 1960) and Chang (1963) where the main problem was the consistency of the axioms of extensionality and comprehension in set theory.

The starting point for the work of Klaua (1965) and the extended version (1967a) is also some set J of urelements over which, inside some basic set theory, a whole cumulative hierarchy $\underline{E}$ (of height ω only, for simplicity) of sets is considered. Inside this universe of discourse $\underline{E}$, a many-valued identity $=_w$, is defined starting with a so-called identity function $\underline{id}$: $J \times J \rightarrow \underline{w}$ with properties:

(i) $\underline{id}(x,y) = 1$ iff $x = y$

(ii) $\underline{id}(x,y) = \underline{id}(y,x)$

(iii) $\underline{id}(x,y) + \underline{id}(y,z) - 1 \leq \underline{id}(x,z)$,

i.e. $\underline{id}$ is chosen in such a way that by

$$\rho(x,y) =_{df} 1 - \underline{id}(x,y)$$

a metric ρ with maximal distance 1 is defined in J. The recursive definition of $=_w$ is such that the author gets for example

$$\models x =_w x \ , \tag{1}$$
$$\models x =_w y \longrightarrow y =_w x \ , \tag{2}$$
$$\models x =_w y \ \& \ y =_w z \longrightarrow x =_w z . \tag{3}$$

The main step for this recursion proceeds in such a way that for sets x,y in the same level inside E there holds true

$$\langle x =_w y \rangle = \langle \forall_x u \exists_y v (u =_w v) \wedge \forall_y v \ \exists_x u (v =_w u) \rangle \ .$$

With this generalized identity $=_w$ the generalized, i.e. many-valued, membership relation ε and inclusion relation $\subseteq_w$ are defined by

$$x \,\varepsilon\, y =_{df} \exists_y v (x =_w v) \ ,$$
$$x \subseteq_w y =_{df} \forall_x u \ \exists_y v (v =_w v) .$$

From these definitions the author then proves (Klaua, 1967, p. 291)

$$\models z =_w x \ \& \ x \varepsilon y \longrightarrow z \,\varepsilon\, y \ , \tag{4}$$
$$\models x \,\varepsilon\, y \ \& \ y \subseteq_w z \longrightarrow x \,\varepsilon\, z \ , \tag{5}$$
$$\models x \subseteq_w y \ \& \ y \subseteq_w z \longrightarrow x \subseteq_w z \ , \tag{6}$$
$$\models x \subseteq_w y \longleftrightarrow \forall u (u \,\varepsilon\, x \longrightarrow u \varepsilon y) \ . \tag{7}$$

He is able to prove a suitable many-valued version of extensionality in full generality, but for comprehension he gets only a limited version.

2.2. Characteristic functions generalized

Perhaps this problem with comprehension cause the second start (1966a) a short time later. Again, a set I of urelements is assumed, and over it a full transfinite cumulative hierarchy of generalized sets is constructed. Each one of these generalized sets now is defined as a generalized characteristic function with values in w.

To explain this construction in more detail, let us call partial universes and denote by $V_\alpha, \alpha \in On$ the successive stages of this construction. $V_o = I$. For limit ordinals $\lambda = 0$, V_λ is the union of all V_α, $\alpha < \lambda$. For successor ordinals $\alpha + 1$, the set $V_{\alpha+1}$ is the union of V_α with the set of all those functions A: $V_\alpha \rightarrow w$ for which there does not exist any ordinal $\beta < \alpha$ and function $\overline{B}$: $V_\beta \rightarrow w$ such that $|A| = |B|$. Here we use $|C|$ for the "support" $\{x \in dom\ C | C(x) \neq 0\}$. The whole universe V of generalized sets is the union of all the partial universes.

Based on this construction, a generalized membership notation $x \varepsilon A$ is introduced, such that

$$\langle x \varepsilon A \rangle = \begin{cases} A(x) & \text{if } x \in \text{dom } A , \\ 0 & \text{if } x \notin \text{dom } A . \end{cases} \tag{8}$$

Using $\bigvee$ for the truth value 1 and $\bigwedge$ for the truth value 0 as logical constants in the language of many-valued logic, the further set theoretical notions of inclusion and equality can be generalized, and they are introduced in Klaua (1966a) as:

$$x \subseteq_w y =_{df} \begin{cases} \forall u (u \varepsilon x \rightarrow u \varepsilon y) & \text{if } x,y \notin \underline{I} \\ \bigwedge & \text{otherwise,} \end{cases} \tag{9}$$

$$x =_w y =_{df} \begin{cases} \bigvee & \text{if } x,y \in \underline{I},\ x=y \\ x \subseteq_w y \wedge y \subseteq_w x & \text{otherwise.} \end{cases} \tag{10}$$

It is now straightforward to prove (1), (2), (3), (5), (6), (7) again. However, this process fails on (4).

Again, by

$$\rho(x,y) = 1 - \langle x =_w y \rangle = \langle \neg x =_w y \rangle$$

a metric in V is given. With respect to this metric some notions of continuity of generalized sets are introduced in (1966a) and these can be used to show that under some additional assumptions (4) holds "approximately" true.

For this second approach also full many-valued versions of extensionality and comprehension hold true. In this sense, the second approach was superior to the first one.

The generality of this second approach was also made clear through the construction of an isomorphic embedding of a given cumulative universe of classical set theory into the universe V, lending weight to the assertion that V is a generalization of the usual cumulative universe of crisp sets (Klaua, 1967b).

Obviously, the generalized "many-valued" sets of the first level defined by this second approach, i.e. the members of $V_1 \setminus V_o$, are the fuzzy sets in $\underline{I}$ in the terminology of Zadeh (1965). And for the case $\underline{w} = [0,1]$, this set $V_1 \setminus V_o$ is exactly the set of all fuzzy sets in $\underline{I}$. Hence, the fuzzy sets of Zadeh (1965) are special cases of the "many-valued" sets of Klaua (1966a), and his universe V is a kind of universe of fuzzy sets closed under fuzzification.
Of course, $\langle x \varepsilon A \rangle$ as defined in (8) is nothing else than Zadeh's $\mu_A(x)$, and for fuzzy sets A,B: $A \subseteq B$ in the sense of Zadeh (1965) holds true iff $\langle A \subseteq_w B \rangle = 1$. Therefore, in the following we use the name "fuzzy set" for all the generalized sets in V, independent of the finiteness or infiniteness of the set $\underline{w}$ of generalized membership, i.e. truth values.

This connection between the two approaches of Klaua (1966a) and Zadeh (1965) was in Klaua's group seen immediately after receiving information concerning the Zadeh paper. This was in 1968. In particular, the paper of Klaua (1966a) was fully independent of Zadeh (1965). Although the first printed references to Zadeh (1965) are only in Klaua (1970, 1972) and Gottwald (1972), the systematic treatment and generalization of Zadeh (1965) along the lines of Klaua (1966a, 1966b) have been realized already in an unpublished master thesis some years earlier (Jahn, 1969).

To avoid misunderstanding it has to be emphasized that Zadeh, from the very beginning, was most interested in applications of fuzzy sets in different fields and this was one of the reasons for his great influence, whereas Klaua and his followers in the first years concentrated wholly on the development of a general set theory for these new, generalized sets.

3. ALGEBRA OF FUZZY SETS, FUZZY RELATIONS AND MAPPINGS

In his work (1966a) Klaua was already defining the union and intersection of his fuzzy sets by the maximum resp. minimum of the corresponding membership values, i.e. by use of the many-valued connectives $\vee, \wedge$ in the same way as Zadeh (1965). And he proves that all the sets $V_\alpha \setminus V_0$ are distributive lattices with suitably definable zeros and ones. With a difference of fuzzy sets defined through

$$u \varepsilon (x - y) =_{df} u \varepsilon x \wedge \neg u \varepsilon y$$

a generalized complementation is also available, again being a generalization of the complementation in Zadeh (1965).

The sequel paper, Klaua (1966b), further develops set algebra for fuzzy sets. A whole family of new complementation operations is introduced, the cartesian product of fuzzy sets defined, and each one of the operations of taking the union, the intersection and the cartesian product of fuzzy sets is generalized from a binary operation to an infinitary one for families of fuzzy sets. In every case, many laws for these operations are proved, which we will not describe in detail here.

Much more of interest is another aspect of this paper that has to be explained: the strict limitation to the finite cases for the truth value set only. At first, this looks like a large difference to the Zadeh's approach.

Of course, from the formal point of view it is a specialization. And really, this way was chosen to avoid some technical difficulties connected with the case $\underline{w} = [0,1]$ and mainly caused by the existence of subsets of $[0,1]$ which do not contain their infimum resp. supremum. However, it was always the conviction - which never was disproved - in accepting this restriction that the remaining cases would be sufficiently representative also for $\underline{w} = [0,1]$.
The paper Jahn (1969) indeed strongly supported this view; and Klaua (1970) sketched the development of the algebra of fuzzy sets in the way of his (1966b) paper for the case $\underline{w} = [0,1]$, excluding the infinitary operations - and, as claimed, all the results of (1970) are as in Klaua (1966b) or a bit more general.

On the other hand, Zadeh and his followers also often restrict the considerations to fuzzy sets with finite support - and therefore with de facto finitely many generalized membership values also. Furthermore, if one is accepting the point of view that an order preserving transformation of the set of generalized membership values does not affect the general laws of fuzzy set theory, all considerations concerning finitely many fuzzy sets with finite support can be done via a suitable order preserving transformation of $[0,1]$ already "isomorphically" in some system of fuzzy sets with a suitable w_n as set of generalized membership values.

One more restriction of this paper - unnecessary from a theoretical point of view, but also simplifying some of the developments - should be discussed here. It concerns the consistent use of many-valued logic begun with definitions (9), (10) of fuzzy inclusion relation and fuzzy equality as really many-valued relations capable of assuming truth values different from 0 and 1. In fact, this method of analysis is not continued. Instead, the author considers the relations for fuzzy sets de facto as binary ones, e.g. he distinguishes for fuzzy inclusion (9) only $\langle x \subseteq_w y \rangle = 1$ and $\langle x \subseteq_w y \rangle \neq 1$. And new notions, e.g. that of being a fuzzy mapping or that of uniqueness of fuzzy mappings, are also introduced solely in this "binary manner". It is only with Klaua (1970) that the more general point of view is adopted again.

In Klaua (1966b) the notion of fuzzy mapping is introduced in such a way that a fuzzy set is a fuzzy mapping if it is a fuzzy subset of some set of ordered pairs. Furthermore, also the notions of inverse mapping, of the product of mappings, of the domain and range of a fuzzy mapping, and of the image $F\langle A\rangle$ of fuzzy set A under fuzzy mapping F are introduced. In every case, again, standard properties of those set theoretic objects are generalized from classical to fuzzy set theory. Remembering the set theoretically obvious fact that "mapping" and "binary relation" are only different names for the same set theoretical objects, the notion of fuzzy relation therefore was defined also. And again it is interesting to note that not only the fuzzy mappings of Klaua are (higher level generalizations of) Zadeh's fuzzy relations, but also both notions of products, i.e. of composition of mappings resp. relations are the same. Moreover, the image of a fuzzy set under a fuzzy mapping is already the conditioned fuzzy set of Bellman/Zadeh (1970), and this is the same as the max.-min. composition of a fuzzy relation with a fuzzy set much used in fuzzy relation equations presently.

As a last point, the uniqueness of fuzzy mappings is discussed. For this two different notions are defined for fuzzy mappings F:

$$F \text{ unique iff } \models [(x,y_1)\, \varepsilon\, F \wedge (x,y_2)\, \varepsilon\, F \longrightarrow y_1 \mathrel{\dot{=}} y_2] \tag{11}$$

$$F \text{ unique}^{*} \text{ iff } \models [u =_w x \wedge (u,v)\, \varepsilon\, F \wedge (x,y)\, \varepsilon\, F \rightarrow v =_w y] \tag{12}$$

with $=_w$ as in (10) and $x \mathrel{\dot{=}} y$ characterized by the two equivalences: $\langle x \mathrel{\dot{=}} y\rangle \neq 0$ iff $x \neq y$, $\langle x \mathrel{\dot{=}} y\rangle = 1$ iff $x = y$. Each one of these notions gives rise to a corresponding notion of equivalence (equipollence) of fuzzy sets.

Some more binary operations for fuzzy sets are introduced in Gottwald (1969) - published in his (1971b) work -, e.g. two differences $-_o, -_w$ and a symmetric difference Δ such that for fuzzy sets A,B: $A -_o B$ is the restriction of the function A to the difference $|A|\setminus|B|$ of supports $|A|$, $|B|$, and such that

$$\models x\, \varepsilon\, A -_w B \longleftrightarrow x \varepsilon A \;\&\; \neg x \varepsilon B ,$$

$$\models x\, \varepsilon\, A \triangle B \longleftrightarrow x \varepsilon A -_o B \wedge x \varepsilon B -_o A .$$

These notions are used for the definition of fields and σ-algebras of fuzzy sets which are needed for the generalization of measure functions (cf. 5.).

Fuzzy relations, particularly ordering relations, are also considered in Gottwald (1974). There the main point is only to demonstrate that the use of the language of many-valued logic gives a much more coherent treatment and makes analogies of fuzzy relations and crisp relations more visible for treatment than goes the non-use of the language of many-valued logic by authors like Zadeh (1971) in this case.

The inherent generality of the constructions in Klaua (1966a, 1966b) with respect to the structure of truth values considered was pointed out also in Schwartz (1972), where the author takes as truth value structure any universal algebra with one nullary and finitely many finitary operations, giving the construction of generalized partial universes V_α à la Klaua (1966a), and he proves that every Horn sentence holding true in truth value structure also holds true in all the $V_{\alpha+1} \setminus V_0$ for the set algebraical operations induced canonically by the finitary operations of truth value structure. A Horn sentence is a formula of predicate calculus in prenex form, whose matrix is a conjunction of disjunctions, each one built up from term equations or their negations, but with at most one unnegated term equation within each disjunction. Of course, the L-fuzzy sets of Goguen (1967) are covered by this general construction, and the R-fuzzy sets of Wechler/Dimitrov (1974) also.

4. FUZZY CARDINALS

The first generalizations of the theory of cardinal numbers to fuzzy sets are in Gottwald (1969), (1971a) and are based on Klaua (1966b). Two definitions of equipollence of fuzzy sets are given by Klaua in the usual way with the corresponding notions of the algebra of fuzzy sets: A,B are equivalent (*) iff A,B are domain resp. range of a fuzzy mapping F which is unique(*) together with its inverse mapping.

For uniqueness in the sense of (11) it proved to be very useful to have the notion of the t-component of a fuzzy set introduced in Klaua (1966b). This notion, defined for any fuzzy set A as

$$A^t = \{ x \mid \langle x \varepsilon A \rangle = t \} = \{ x \mid \mu_A(x) = t \},$$

can be used in the same way for representing a fuzzy set A by the family $(A^t \mid 0 \neq t \in \underline{w})$ of crisp sets as this is also the case with the corresponding t-cuts

$$A_t = \{ x \mid \langle x \varepsilon A \rangle > t \} = \{ x \, \mu_A(x) > t \}$$

used by Zadeh (1965). Both of these notions sometimes are very useful for characterization of properties of fuzzy sets. And it depends of the special property to consider, which one of these notions is the more suitable one: convexity of a fuzzy set A in sense of Zadeh (1965) can easily be characterized by the usual convexity of all its t-cuts for $t \neq 0$, and equipollence of fuzzy sets A,B based on definition (11) of uniqueness correspondingly can easily be characterized by the usual equipollence of all pairs of corresponding t-components A^t, B^t for $t \neq 0$ (Klaua, 1966b).

With the set algebraic notions for fuzzy sets discussed up to this point, it was possible to generalize the usual set theoretic construction of cardinals as suitable classes of equipollent sets in a straightforward way to fuzzy sets, and to do so also for the usual definitions of sum and product of cardinals and for the well-ordering of the cardinals. In this way, two notions of the cardinal number of fuzzy set arose, based on definition (11) resp. (12) of uniqueness. Let us refer to them as cardinals of first resp. second kind. Each generalized cardinal number of a fuzzy set itself is a crisp set of fuzzy sets.

From the main point of difference in (11), (12), namely the essential use of many-valued equality $=_w$ for (12), the result is that the quotient structure of the algebraic structure of all cardinals of the second kind, with respect to the operations of sum and product and to the ordering relation, by an appropriate congruence relation is isomorphic to the algebraic structure of all cardinals of first kind. In this sense, with cardinals of second kind a finer distinction of fuzzy sets is possible as with cardinals of first kind.

For the fuzzy sets of Zadeh, i.e. for the fuzzy sets of first level in the whole cumulative universe of fuzzy sets, both kinds of cardinals of fuzzy sets coincide.

Cardinals of first kind are much easier to treat because for them a representation is possible as a family of cardinals of crisp sets: given a fuzzy set A, its cardinal $card_1 A$ of first kind has as representation the family $(\underline{a}_t \mid 0 \neq t \in \underline{w})$ of the cardinals $\underline{a}_t$ of the t-components A^t of A. And, as proved in Gottwald (1969), (1971a), this representation can also be used to give characterizations of the sum, the product and the ordering relation for cardinals of the first kind of fuzzy sets.

For given fuzzy sets A,B with $\underline{a} = card_1 A$ represented by $(\underline{a}_t \mid 0 \neq t \in \underline{w})$ and $\underline{b} = card_1 B$ represented by

$(\underline{b}_t \mid 0 \neq t \in \underline{w})$ it holds true

$$\underline{a} \leqslant \underline{b} \quad \text{iff} \quad \underline{a}_t \leqslant \underline{b}_t \quad \text{for all} \quad 0 \neq t \in \underline{w}$$

and the representations of sum and product are given for

$$\underline{a} + \underline{b} \quad \text{by} \quad (\underline{a}_t + \underline{b}_t \mid 0 \neq t \in \underline{w}) \ ,$$

$$\underline{a} \cdot \underline{b} \quad \text{by} \quad (\underline{a}_t \underline{b}_t + \underline{a}_t \underline{B}(t) + \underline{b}_t \underline{A}(t) \mid 0 \neq t \in \underline{w})$$

with

$$\underline{A}(t) = \sum_{t < s \in \underline{w}} \underline{a}_s \ , \qquad \underline{B}(t) = \sum_{t < s \in \underline{w}} \underline{b}_s \ .$$

The position of Klaua (1966b), to consider only finite sets $\underline{w}$ of generalized membership, i.e. truth values, is also adopted in Gottwald (1969), (1971a,b), but the same kind of representation works as well for $w = [0,1]$. Care is needed only in using the sums $\underline{A}(t)$, $\underline{B}(t)$ for the representation of the product, yet only for fuzzy sets with infinite support.

As a side remark, let us also mention that a notion of the well-ordering of a fuzzy set is defined in Gottwald (1969), (1971a) in such a way that a well-ordering of fuzzy set A induces an usual well-ordering in every t-component ($t \neq 0$) of A and makes elements of different components of A incomparable. On this basis, ordinals of well-ordered fuzzy sets are defined and represented by families of usual ordinals in essentially the same manner as for cardinals of the first kind.

Via the representation of card_1 A or via the support of fuzzy set A one gets a definition of finiteness of fuzzy sets and cardinals of fuzzy sets, and hence also a generalized notion of "fuzzy" integer. But, those old "fuzzy" numbers should not be confused with the fuzzy numbers of present day theory of fuzziness, which are fuzzy subsets of, for example, the set of real numbers.

It is interesting to note that these "fuzzy" integers built up a structure in which every element has exactly as many different immediate successors as $\underline{w}$ has elements different from zero. Additionally, in this structure suitable generalized versions of the Peano axioms hold true (Gottwald, 1972). Hence, the structure of the "fuzzy" integers has some analogies with the partially ordered arithmetics of Vučković (1959). And it is also not difficult to determine all the isomorphic embeddings of the (usual) integers into these structures (Gottwald, 1973).

Another approach to a theory of cardinals in fuzzy set theory is realized in Klaua (1970, 1972, 1973). It is based on a more consequent use of many-valued notions as in his (1966b) and really produces fuzzy sets as generalized cardinals.

The starting point is the definition of the notion "F is a bijection from A onto B" for fuzzy sets A,B and fuzzy mappings F as a many valued notion. These definitions are different in Klaua (1970) and (1972). The paper (1970) starts with many-valued uniqueness à la (11)

$$\mathrm{Un}(F) =_{df} (\forall)\ (x =_w u \wedge (x,y) \varepsilon F \wedge (u,v) \varepsilon F \longrightarrow \longrightarrow y =_w v),$$

using $(\forall)$ as many-valued generalization of all free variables, and defines with F^{-1} for the inverse fuzzy mapping of F, f-dom resp. f-rg for the fuzzy domain and fuzzy range of fuzzy mapping

$$\mathrm{Bij}\ (F,A,B) =_{df} (\mathrm{Un}(F) \wedge \mathrm{Un}(F^{-1}))\ \&\ \text{f-dom } F \subseteq_w A\ \&\ A \subseteq_w \text{f-dom } F\ \&\ \text{f-rg } F \subseteq_w B\ \&\ B \subseteq_w \text{f-rg } F\ .$$

In Klaua (1972) for fuzzy sets A,B and crist (!) 1 - 1 surjections f of the supports $|A|$, $|B|$ the definition is

$$\mathrm{Bij}'(f,A,B) =_{df} \forall_{\mathrm{dom}(f)} x(x \varepsilon A \longleftrightarrow f(x) \varepsilon B)\ .$$

If we define with Klaua (1970) a fuzzy equivalence relation in a class X as such a binary fuzzy relation R in $X \times X$ with

$$\models \forall_X x(xRx)\ ,$$

$$\models \forall x \forall y(xRy \rightarrow yRx)\ ,$$

$$\models \forall x \forall y \forall z(xRy\ \&\ yRz \rightarrow xRz)\ ,$$

where we used the notation "xRy" instead of "$(x,y) \varepsilon R$", then for each element a of X its fuzzy equivalence class $[a]$ with respect to R is the fuzzy set characterized by

$$\models x \varepsilon [a] \longleftrightarrow xRa\ .$$

Now, defining equipollence for fuzzy sets as that fuzzy relation which has for any fuzzy sets A,B as truth value the supremum of all the value $\langle \mathrm{Bij}(F,A,B)\rangle$ resp. $\langle \mathrm{Bij}'(f,A,B)\rangle$ for all admissible F,f, then fuzzy equipollence of fuzzy sets is a fuzzy equivalence relation in the class of all fuzzy sets. And the fuzzy cardinals are the fuzzy equivalence classes of fuzzy equipollence.

Following the usual procedure in set theory, Klaua (1972, 1973) defined also the sum and product of fuzzy cardinals, a fuzzy ordering relation of fuzzy cardinals, and the finiteness of fuzzy sets (via finiteness of support) and fuzzy integers as fuzzy cardinals of finite fuzzy sets.

In every case, the main properties of these new notions are studied; and stress is laid in these papers on continuity properties of the new notions with respect to the metric given by negated fuzzy equality.

5. GENERALIZED REAL NUMBERS AND FUZZY MEASURES

The way of constructing cardinals of fuzzy sets chosen in Gottwald (1969) and described in the last section could be extended to a generalization of the usual set theoretic way of construction of real numbers starting from the integers. This was done in his (1969, 1971a) works starting with the finite cardinals of first kind. For the cardinals of second kind that approach was not carried through because no interesting non-trivial further results could be proved or conjectured.

For integers of first kind, and also for the generalized "fuzzy" rationals and "fuzzy" reals of this construction, always the representability by a suitable family of usual integers, usual rationals, usual reals was proved. The formulas describing via representation sum, product and ordering for cardinals of first kind did this job also for the generalized "fuzzy" rationals and "fuzzy" reals. The algebraic structures of these sets of "fuzzy" rationals resp. reals with respect to addition and multiplication proved to be commutative rings with zero and one, but with divisors of zero and nontrivial units also.

Yet, these divisors of zero could be characterized through their representations: any "fuzzy" real with representation $(\underline{r}_t \mid 0 \neq t \in \underline{w})$ is a divisor of zero if there exist some $s \in \underline{w}$, $s \neq 0$ such that the sum of all $\underline{r}_t$ for $t \geqslant s$ is zero.

Excluding the divisors of zero, the remaining "fuzzy" rationals resp. reals built up fields with respect to addition and multiplication.

Using the so constructed "fuzzy" reals, Gottwald (1969, 1971b) generalized the introductory notions of the theory of finitely and infinitely additive measures. He is considering as domains of those measures fields and σ-algebras of fuzzy sets; and the ranges of those measures are subsets of fuzzy sets; and the ranges of those measures are subsets of the set of all "fuzzy" reals which are greater or equal to the "fuzzy" real number zero.

With many details, the theory of these functions is developed. Regularity, inner and outer measures, relativization, kernel and hull of equal measure, completeness, and product measures are defined (sometimes of several kinds) and studied with their properties and the relationships between them.

Of course, this notion of "fuzzy" measure is completely different from the measures of fuzziness introduced by deLuca/Termini (1972) and also the fuzzy measures of Sugeno (1974).

6. THE 3-VALUED CASE

The whole interpretation of fuzzy sets has some intrinsic 3-valuedness: for a fuzzy set A which describes some property P,

- $\langle b \varepsilon A \rangle = 1$ means that b surely has property P,
- $\langle b \varepsilon A \rangle = 0$ means that b surely does not have property P, and
- all the other cases for $\langle b \varepsilon A \rangle$ represent some "uncertainty" anout b's having of property P.

Furthermore, if considering the finite cases for $\underline{w}$, it is also the simplest case to have only $\underline{w} = \{0, 0.5, \overline{1}\}$.

This simplest case was treated separately in Klaua (1968, 1969a, 1969b). The representation by the family of t-cuts now is a representation of such a 3-valued fuzzy set A by an ordered pair (A_+, A_{++}) such that

$$x \in A_{++} \text{ iff } \langle x \varepsilon A \rangle = 1 ,$$
$$x \in A_{+} \text{ iff } \langle x \varepsilon A \rangle \geqslant 0.5 .$$

The ordered pairs (A_+, A_{++}), subject only to $A_{++} \subseteq A_+$, are in Klaua (1968) identified with the 3-valued fuzzy sets and named "partial sets".

Obviously, the "flou sets" of Negoita/Ralescu (1975) are exactly the same objects. And the partial sets can also be identified with intervals of crisp sets: (A_+, A_{++}) with the class of all X with $A_{++} \subseteq X \subseteq A_+$.

For the partial sets, Klaua (1968) introduces the most central notions of set algebra like union, intersection, power set, relative complement, cartesian product, mapping, and equivalence relation. The sequel paper (1969a) continues this study and generalizes it in such a way that most notions and relations of set algebra for partial sets - not the operations - are introduced in three different, but strongly related ways, reflecting in doing so the many-valuedness of such notions which is made more explicit for the general case later on in Klaua (1970).

There is an uniform principle for the construction of the three generalizations $\underline{R}_+, \underline{R}_{++}, \underline{R}_*$ of a relation R for crisp sets to partial sets. To describe it, let us consider, for the partial set $A = (A_+, A_{++})$, its interval

$$\text{int } A = \{ X \mid A_{++} \subseteq X \subseteq A_+ \} .$$

Then the definitions of $\underline{R}_+, \underline{R}_{++}, \underline{R}_*$ are for a given binary relation R between crisp sets for any partial sets A,B:

$$\begin{array}{lll} A \; \underline{R}_+ \; B & \text{iff} & (\forall A' \in \text{int } A)(\forall B' \in \text{int } B)(A'RB') \quad , \\ A \; \underline{R}_{++} B & \text{iff} & (\exists A' \in \text{int } A)(\exists B' \in \text{int } B)(A'RB') \quad , \\ A \; \underline{R}_* \; B & \text{iff} & (\forall A' \in \text{int } A)(\exists B' \in \text{int } B)(A'RB') \\ & & \text{and } (\forall B' \in \text{int } B)(\exists A' \in \text{int} A)(A'RB') \quad , \end{array}$$

where quantifications are that of classical two-valued logic. For unary R these definitions are changed to

$$\begin{array}{lll} \underline{R}_+ \; (A) & \text{iff} & (\forall A' \in \text{int } A)(R(A')) \quad , \\ \underline{R}_{++}(A) & \text{iff} & (\exists A' \in \text{int } A)(R(A')) \quad , \\ \underline{R}_* \; (A) & \text{iff} & R(A_+) \text{ and } R(A_{++}) \quad . \end{array}$$

One of the three relations of equipollence of partial sets of Klaua (1969a), $\sim_*$, is especially suitable for the introduction of "partial cardinals". The main reason for this is, as in Gottwald (1969), the possibility of representation by pairs of usual cardinals. Set theoretic definition of sum and product of partial cardinals, and the set theoretic construction of "partial" real numbers starting with finite partial cardinals, gives, via representation as, resulting set of "partial" reals, exactly the set of interval numbers of interval mathematics.

That result is mentioned in Klaua (1969b). And then this paper is devoted to a generalization of interval arithmetics: the main relations of interval arithmetics, i.e. identity and ordering of interval numbers, are introduced à la Klaua (1969a) in three ways.

This "3-valued" look at interval mathematics - which does not consist in really using the language of 3-valued logic - is continued in other papers (e.g. Jahn, 1974, 1975a) but outside the scope of this paper. If a corresponding point of view could be useful also for the arithmetic of fuzzy numbers (as fuzzy subsets of the real line) is an open problem.

Yet, a by-product of this development deserves to be mentioned. Jahn (1975b) develops a system of generalized fuzzy sets up to set algebra, where the membership values of his generalized fuzzy sets are intervals of the truth value set $\underline{w} = [0,1]$. Therefore, his so called "interval valued sets" are special fuzzy sets of type 2 of Zadeh (1975). But, this system was not developed further.

7. CONCLUDING REMARKS

The further work on fuzziness continuing the researches surveyed here has been published mainly in english language papers, e.g. in Fuzzy Sets and Systems and other related sources well known to the fuzzy public. Hence, it is not necessary to review it also. The interested reader can also consult the survey paper by Gottwald (1981).

Karl Marx University, Leipzig, G.D.R.

REFERENCES

Bellman, R.E. and L.A. Zadeh: 1970, 'Decision making in a fuzzy environment', Management Sci. 17, 141-164.

Chang, C.C.: 1963, 'The axiom of comprehension in infinite valued logic', Mathematica Scand. 13, 9-30.

deLuca, A. and S. Termini: 1972, 'A definition of a non-probabilistic entropy in the setting of fuzzy sets theory', Inform. and Control 20, 301-312.

Giles, R.: 1976, 'Lukasiewicz logic and fuzzy set theory', Internat. J. Man-Machine Stud. 8, 313-327.

Goguen, J.A.: 1967, 'L-fuzzy sets', J. Math. Anal. Appl. 18, 145-174.

Gottwald, S.: 1969, Konstruktion von Zahlbereichen und die Grundlagen der Inhaltstheorie in einer mehrwertigen Mengenlehre, Ph.D. Thesis, Leipzig, Karl Marx University.

Gottwald, S.: 1971a, 'Zahlbereichskonstruktionen in einer mehrwertigen Mengenlehre', Zeitschr. math. Logik Grundlagen Math. 17, 145-188.

Gottwald, S.: 1971b, 'Elementare Inhalts- und Masstheorie in einer mehrwertigen Mengenlehre', _Mathematische Nachr_. 50, 27-68.

Gottwald, S.: 1972, 'Verallgemeinerte Peano-Systeme', _Zeitschr. math. Logik Grundlagen Math_. 18, 19-30.

Gottwald, S.: 1973, 'Uber Einbettungen in Zahlenbereiche einer mehrwertigen Mengenlehre', _Mathematische Nachr_. 56, 43-46.

Gottwald, S.: 1974, 'Mehrwertige Anordnungsrelationen in klassischen Mengen', _ibid_. 63, 205-212.

Gottwald, S.: 1981, 'Fuzzy-Mengen und ihre Anwendungen. Ein Uberblick', _Elektron. Informationsverarb. Kybernetik_ EIK 17, 207-235.

Jahn, K.U.: 1969, _Uber einen Ansatz zur mehrwertigen Mengenlehre unter Zulassung kontinuum-vieler Wahrheitswerte_, Master Thesis, Leipzig, Karl Marx University.

Jahn, K.U.: 1974, 'Eine Theorie der Gleichungssysteme mit Intervall-Koeffizienten', _ZAMM_ 54, 405-412.

Jahn, K.U.: 1975a, 'Eine auf den Intervall-Zahlen fussende 3-wertige lineare Algebra', _Mathematische Nachr_. 65, 105-116.

Jahn, K.U.: 1975b, 'Intervall-wertigen Mengen', _ibid_. 68, 115-132.

Klaua, D.: 1965, 'Über einen Ansatz zur mehrwertigen Mengenlehre', _Monatsber. Deut. Akad. Wiss. Berlin_ 7, 859-876.

Klaua, D.: 1966a, 'Über einen zweiten Ansatz zur mehrwertigen Mengenlehre', _ibid_. 8, 161-177.

Klaua, D.: 1966b, 'Grundbegriffe einer mehrwertigen Mengenlehre', _ibid_. 8, 782-802.

Klaua, D.: 1967a, 'Ein Ansatz zur mehrwertigen Mengenlehre', _Mathematische Nachr_. 33, 273-296.

Klaua, D.: 1967b, 'Einbettung der klassischen Mengenlehre in die mehrwertige', _Monatsber. Deut. Akad. Wiss. Berlin_ 9, 258-272.

Klaua, D.: 1968, 'Partiell definierte Mengen', _ibid_. 10, 571-578.

Klaua, D.: 1969a, 'Partielle Mengen mit mehrwertigen Grundbeziehungen', _ibid_. 11, 573-584.

Klaua, D.: 1969b, 'Partielle Mengen und Zahlen', _ibid_. 11, 585-599.

Klaua, D.: 1970, 'Stetige Gleichmächtigkeiten kontinuierlich-wertiger Mengen', _ibid_. 12, 749-758.

Klaua, D.: 1972, 'Zum Kardinalzahlbegriff in der mehrwertigen Mengenlehre', in: G. Asser et al. (eds.), _Theory_ *of Sets and Topology*, Deutscher Verlag d. Wiss., Berlin, pp. 313-325.

Klaua, D.: 1973, 'Zur Arithmetik mehrwertiger Zahlen', *Mathematische Nachr*. 57, 275-306.

Negoita, C.V. and D.A. Ralescu: 1975, *Applications of Fuzzy Sets to Systems Analysis*, Birkhäuser-Verlag, Basel.

Schwartz, D.: 1972, 'Mengenlehre über vorgegebenen algebraischen Systemen', *Mathematische Nachr*. 53, 365-370.

Skolem, Th.: 1957, 'Bemerkungen zum Komprehensionsaxiom', *Zeitschr. math. Logik Grundlagen Math*. 3, 1-17.

Skolem, Th.: 1960, 'A set theory based on a certain 3-valued logic', *Mathematica Scand*. 8, 127-136.

Sugeno, M.: 1974, *Theory of Fuzzy Integrals and its Application*, Ph.D. Thesis, Tokyo, Tokyo Institute of Technology.

Vucković, V.: 1959, 'Partially ordered recursive arithmetics', *Mathematica Scand*. 7, 305-320.

Wechler, W. and V. Dimitrov: 1974, 'R-fuzzy automata', in: J.L. Rosenfeld (ed.), *Information Processing 1974* North-Holland Publ. Comp., Amsterdam, pp. 657-660.

Zadeh, L.A.: 1965, 'Fuzzy sets', *Inform. and Control* 8, 338-353.

Zadeh, L.A.: 1971, 'Similarity relations and fuzzy orderings', *Information Sci*. 3, 177-200.

Zadeh, L.A.: 1975, 'The concept of a linguistic variable and its application to approximate reasoning - I', *ibid*. 8, 199-250.

Ulrich Höhle and Erich Peter Klement

PLAUSIBILITY MEASURES - A GENERAL FRAMEWORK FOR POSSIBILITY AND FUZZY PROBABILITY MEASURES

INTRODUCTION

The question of how to describe uncertainty and how to deal with it in a rigorous mathematical way has been discussed for many years (we only recall the dispute between objectivists and subjectivists in probability theory). More recently two attempts in the context of fuzzy sets (Zadeh [24]) have been made. In 1968 Zadeh [25] introduced fuzzy probability which led to an axiomatic study of this topic (Höhle [6], Klement [13,15], Klement et al. [17]) in the sense of a valuation on a lattice, the so-called fuzzy σ-algebra (Klement [12,14]), which also might be assumed to be non-distributive. Using a suitable complement terms such as discernibless and indiscernibless may be formulated (Höhle [9]). The second approach, again given by Zadeh [26] in 1978, was the introduction of possibility measures, mappings on a Boolean algebra where the addition is performed using the maximum instead of the sum.

A very general framework for these two concepts is provided by the plausibility measures going back to the work of Shafer [22,23].
The leading idea for our studies was now to characterize these new mathematical objects by well-known ones such as ordinary probability measures and Markov kernels in order to have powerful mathematical machinery at hand when working with possibility and fuzzy probability measures. This characterization has been done by Höhle [7] in general for plausibility measures and in particular by Klement [13,15] for a large class of fuzzy probability measures. An important tool in doing so was the idea of realizations of fuzzy σ-algebras (Łos [19,20]). An immediate application of these studies could be the development of a probability theory on ortholattices (see example 1.1).

The organization of this paper is as follows: first we

H. J. Skala, S. Termini, and E. Trillas (eds.), Aspects of Vagueness, 31–50.

define fuzzy σ-algebras and introduce the important operations which are derived from triangular norms (section I); in the next section we give the axioms for plausibility measures, fuzzy probability measures and possibility measures; in section III we discuss realizations of a fuzzy σ-algebra and characterize plausibility measures by canonical Borel probability measures associated with them; section IV brings a classification of fuzzy probability measures using Markov kernels; in the subsequent section we consider entropies of discernibleness and indiscernibleness and calculate them in many important cases; finally we discuss some possible fuzzifications of these concepts.

I. ⊗-FUZZY σ-ALGEBRAS

Let $(L,\leq,\mathbf{0},\mathbf{1})$ be a σ-complete lattice containing a smallest element $\mathbf{0}$ and a greatest element $\mathbf{1}$, let $c : L \to L$ be a mapping and let $\otimes$ be a binary operation on L such that $\mathbf{1}$ acts as a unit and $\mathbf{0}$ as a zero element of L.
Then we call the quadruple $(L,\otimes,\leq,c)$ a ⊗-fuzzy σ-algebra if the following conditions are satisfied:

[FA1] $(L,\otimes,\leq)$ is a partially ordered semigroup (see Fuchs [5]);
[FA2] $\forall\ \mu,\upsilon \in L$: $(\mu \vee \upsilon) \otimes (\mu \wedge \upsilon) = \mu \otimes \upsilon$;
[FA3] c is an order reversing involution.

It is readily seen that the structure of ⊗-fuzzy σ-algebras is preserved under direct sums.
On a ⊗-fuzzy σ-algebra L we may consider another binary operation ⊕ defined by

$$\mu \oplus \upsilon = c(c(\mu) \otimes c(\upsilon)) \tag{1.1}$$

for all $\mu,\upsilon \in L$. Obviously ⊕ is the dual operation of ⊗ with respect to c.

1.1 Example. Each σ-complete lattice $(L,\leq,\mathbf{0},\mathbf{1})$ equipped with an order reversing involution c is a ∧-fuzzy σ-algebra. Moreover, each σ-complete ortholattice is a ∧-fuzzy σ-algebra. Hence, for instant, the lattice $L(\mathcal{H})$ of all closed subspaces of a Hilbert space $\mathcal{H}$ is a ∩-fuzzy σ-algebra (see Birkhoff [3]).

Usually we shall call ∧-fuzzy σ-algebras simply fuzzy σ-algebras. On the unit interval [0,1], however, the so-called triangular norms are interesting binary operations which have been studied by Schweizer and Sklar [21] in the context of probabilistic metric spaces. A triangular norm (or briefly t-norm) is a mapping $T : [0,1]\times[0,1] \to [0,1]$ fulfilling the following properties:

[TN1] $\forall\ \alpha \in [0,1]:\ T(1,\alpha) = \alpha;$ (boundary condition)
[TN2] $\forall\ \alpha,\beta,\gamma,\delta \in [0,1]:\ \alpha \leq \gamma, \beta \leq \delta \Rightarrow T(\alpha,\beta) \leq T(\gamma,\delta);$ (monotonicity)
[TN3] $\forall\ \alpha,\beta \in [0,1]:\ T(\alpha,\beta) = T(\beta,\alpha);$ (commutativity)
[TN4] $\forall\ \alpha,\beta,\gamma \in [0,1]:\ T(T(\alpha,\beta),\gamma) = T(\alpha,T(\beta,\gamma)).$ (associativity)

Obviously $([0,1],T,\leq)$ is a partially ordered semigroup. The mapping $S : [0,1]\times[0,1] \to [0,1]$ given by

$$S(\alpha,\beta) = 1 - T(1-\alpha,1-\beta) \qquad \forall\ \alpha,\beta \in [0,1] \tag{1.2}$$

is called a t-conorm or the dual of T. It clearly fulfills [TN2], [TN3], [TN4] and

[SN1] $\forall\ \alpha \in [0,1]:\ S(0,\alpha) = \alpha.$ (boundary condition)

The most important t-norms are given by

$$T_0(\alpha,\beta) = \min(\alpha,\beta), \tag{1.3}$$

$$T_1(\alpha,\beta) = \alpha.\beta\ , \tag{1.4}$$

$$T_\infty(\alpha,\beta) = \max(\alpha+\beta-1,\ 0), \tag{1.5}$$

$$T_W(\alpha,\beta) = \begin{cases} \min(\alpha,\beta) & \text{if } \max(\alpha,\beta) = 1\,, \\ 0 & \text{else.} \end{cases} \tag{1.6}$$

It is immediately seen that T_W is the smallest and T_0 is the largest t-norm, i.e. each t-norm T fulfills

$$T_W \leq T \leq T_0. \tag{1.7}$$

For each $s \in\ ]0,1[\cup]1,+\infty[$ we also consider the t-norms T_s defined by

$$T_s(\alpha,\beta) = {}^{s}\log\left(1 + \frac{(s^\alpha-1)(s^\beta-1)}{s-1}\right) \tag{1.8}$$

for all $\alpha,\beta \in [0,1]$. The t-norms T_s, $s \in [0,+\infty]$, form a single family in the sense that the T_s converge pointwise to T_0, T_1 and T_∞ according to the subscripts. In a remarkable paper Frank [4] proved that (T_s,S_s), $s \in [0,+\infty]$, together with their so-called ordinal sums are the only pairs of t-norms and t-conorms fulfilling the functional equation

$$T(\alpha,\beta) + S(\alpha,\beta) = \alpha + \beta. \tag{1.9}$$

1.2 Example. (1) If we equip $[0,1]$ with the usual order reversing involution c given by

$$c(\alpha) = 1-\alpha \qquad \forall\, \alpha \in [0,1] \tag{1.10}$$

then for each t-norm T the quadruple $([0,1],T,\leq,c)$ is a T-fuzzy σ-algebra.

(2) Let T be a t-norm, X an arbitrary non-empty set and $\leq$ the product ordering on $[0,1]^X$. In the usual way we introduce a binary operation and an order reserving involution on the distributive lattice $([0,1]^X,\leq)$ given by

$$T(\mu,\upsilon)(x) = T(\mu(x),\upsilon(x)) \qquad \forall\, x \in X, \tag{1.11}$$
$$(1-\mu)(x) = 1 - \mu(x) \qquad \forall\, x \in X \tag{1.12}$$

for all $\mu,\upsilon \in [0,1]^X$. Now each σ-complete regular sublattice L of $[0,1]^X$ fulfilling

[A1] $\forall\, \mu,\upsilon \in L$: $T(\mu,\upsilon) \in L$;
[A2] $\forall\, \mu \in L$: $1-\mu \in L$;
[A3] $\forall\, \alpha \in [0,1]$: $\alpha.1_X \in L$

is a T-fuzzy σ-algebra (see Klement [14]). As usual, 1_A stands here for the characteristic function of a subset A of X.

(3) If $(X,\mathcal{A})$ is a measurable space let $\zeta(\mathcal{A})$ denote the set of all measurable fuzzy subsets of X, that means all measurable mappings $\mu : X \to [0,1]$ (see Zadeh [25], Klement [12]). Keeping the notation of the previous example for each measurable t-norm T the distributive σ-complete lattice $\zeta(\mathcal{A})$ is a T-fuzzy σ-algebra. We shall call $\zeta(\mathcal{A})$ the (T-)fuzzy σ-algebra generated by $\mathcal{A}$.

The following theorem demonstrates the important role of the t-norms T_s :

1.3 Theorem. Let s be an element of $]0,+\infty]$, X a non-empty set and L be a σ-complete sublattice of $[0,1]^X$ satisfying condition [A3]. Then the following assertions are equivalent:

(i) L is a T_s-fuzzy σ-algebra;
(ii) L is generated, i.e. there exists an ordinary σ-algebra $\mathcal{A}$ on X such that $L = \zeta(\mathcal{A})$.

Proof. Implication (ii) => (i) follows from example 1.2(3). In order to verify (i) => (ii) we first show that $\mu_o, \upsilon_o \in L$ always implies $S_\infty(\mu_o,\upsilon_o) \in L$. This can be done defining monotone sequences $(\mu_n)_{n\in\mathbb{N}}$, $(\upsilon_n)_{n\in\mathbb{N}}$ in L recursively by

$$\mu_n = S_s(\mu_{n-1},\upsilon_{n-1}), \qquad \upsilon_n = T_s(\mu_{n-1},\upsilon_{n-1}). \qquad (n \in \mathbb{N})$$

Since T_s and its dual S_s fulfill the functional equation (1.9) we get $\mu_n + \upsilon_n = \mu_o + \upsilon_o$ for all $n \in \mathbb{N}$. Now using a property of T_s (see Klement [14]) we obtain $\lim_{n\to\infty} \mu_n = S_\infty(\mu_o,\upsilon_o)$. Therewith we can conclude $S_\infty(\mu_o,\upsilon_o) = S_s(\mu_o,\upsilon) \in L$ where υ is given by $\upsilon = \lim_{n\to\infty} S_s(\upsilon_o,\upsilon_1,\ldots,\upsilon_n)$. Now choose $(\mu,\alpha) \in L\times[0,1]$ and put $\pi_n = S_\infty(\mu,(1-\alpha).1_X)$ for all $n \in \mathbb{N}$. Then we obtain $1_{\{\mu\geq\alpha\}} = \lim_{n\to\infty} T_s(\pi_1,\ldots,\pi_n) \in L$. Now it is straightforward that $\mathcal{A} = \{A \subset X \mid 1_A \in L\}$ forms a σ-algebra on X generating L.

II. AXIOMS OF PLAUSIBILITY MEASURES

Plausibility measures have been studied first by Shafer [22]. We shall provide a general setting for them such that fuzzy probability measures (Zadeh [25]) and possibility measures (Zadeh [26]) are contained as special cases.

Let $(L,\otimes,\leq,c)$ be a $\otimes$-fuzzy σ-algebra and $m : L \to [0,1]$ a mapping provided with the following properties:

[M1] $m(\mathbf{0}) = 0, \quad m(\mathbf{1}) = 1,$ (<u>boundary conditions</u>)

[M2] $\forall\ \mu,\upsilon \in L: \quad \mu \leq \upsilon \Rightarrow m(\mu) \leq m(\upsilon),$ (<u>monotonicity</u>)

[M3] $\forall\ (\mu_n)_{n\in\mathbb{N}} \in L^{\mathbb{N}},\ (\mu_n)\uparrow : \ m(\bigvee_{n\in\mathbb{N}} \mu_n) = \sup_{n\in\mathbb{N}} m(\mu_n).$ (<u>σ-smoothness</u>)

We say that m is a <u>plausibility measure</u> on L if m is subadditive, i.e. for all $\mu_1,\ldots,\mu_n \in L$ we have

[PL] $$m(\bigwedge_{i=1}^{n} \mu_i) \leq \sum_{i=1}^{n} (-1)^{i-1} \sum_{1\leq j_1<\ldots<j_i\leq n} m(\bigvee_{k=1}^{i} \mu_{j_k}).$$

We shall call m a <u>$\otimes$-fuzzy probability measure</u> on L if it fulfills for all $\mu,\upsilon \in L$

[PR] $m(\mu \otimes \upsilon) + m(\mu \oplus \upsilon) = m(\mu) + m(\upsilon).$

Finally, m is called a <u>⊗-possibility measure</u> on L if it satisfies for all $\mu, \upsilon \in L$

[PO] $m(\mu \oplus \upsilon) = \max(m(\mu), m(\upsilon))$.

It is readily seen from these definitions that both the set of all plausibility measures and the set of all ⊗-fuzzy probability measures on L are convex.

The following proposition demonstrates that the families of ∧-fuzzy probability measures and ∧-possibility measures are the largest possible ones:

<u>2.1 Proposition</u>. (i) Each ⊗-fuzzy probability measure on L is a ∧-fuzzy probability measure on L.

(ii) Each ⊗-possibility measure on L is a ∧-possibility measure on L.

Proof. (i): Using [PR] and [FA2] we get for all $\mu, \upsilon \in L$

$$\begin{aligned} & m(\mu \vee \upsilon) + m(\mu \wedge \upsilon) = \\ &= m((\mu \vee \upsilon) \otimes (\mu \wedge \upsilon)) + m((\mu \vee \upsilon) \oplus (\mu \wedge \upsilon)) = \\ &= m(\mu \otimes \upsilon) + m(\mu \oplus \upsilon) = m(\mu) + m(\upsilon). \end{aligned}$$

(ii): By virtue of [M2],[PO] and [FA2] we obtain for all $\mu, \upsilon \in L$

$$\begin{aligned} & m(\mu \vee \upsilon) = \max(m(\mu \vee \upsilon), m(\mu \wedge \upsilon)) = \\ &= m((\mu \vee \upsilon) \oplus (\mu \wedge \upsilon)) = \\ &= m(\mu \oplus \upsilon) = \max(m(\mu), m(\upsilon)). \end{aligned}$$

With this result it is now straightforward to identify ⊗-possibility measures and ⊗-fuzzy probability measures as special cases of plausibility measures.

<u>2.2 Corollary</u>. (i) Each ⊗-possibility measure on L is a plausibility measure on L.

(ii) If $(L, \otimes, \leq, c)$ is a <u>distributive</u> ⊗-fuzzy σ-algebra (i.e. if $(L, \leq)$ is a distributive lattice and $(L, \otimes, \leq, c)$ is a ⊗-fuzzy σ-algebra) then each ⊗-fuzzy probability measure on L is a plausibility measure on L.

<u>2.3 Example</u>. (1) Let $(X, \mathcal{A}, P)$ be an ordinary probability space, $\zeta(\mathcal{A})$ the fuzzy σ-algebra generated by $\mathcal{A}$ and let T be a measurable t-norm and S its dual such that the functional equation (1.9) is satisfied. Then P induces a T-fuzzy probability measure m_P on $\zeta(\mathcal{A})$ via (Zadeh [25])

$$m_P(\mu) = \int_X \mu \, dP \qquad \forall \mu \in \zeta(\mathcal{A}). \tag{2.1}$$

(2) Let X be a non-empty set and f be a normal fuzzy subset of X (i.e. a mapping $f : X \to [0,1]$ such that $\sup\{f(x) \mid x \in X\} = 1$). Then f induces a $\cap$-possibility measure m_f on $L = \mathcal{P}(X)$, the power set of X, as follows (see Zadeh [26])

$$m_f(A) = \sup\{f(x) \mid x \in A\} \qquad \forall\ A \subset X. \tag{2.2}$$

(3) Each ordinary probability measure P on a Boolean σ-algebra $\mathcal{A}$ is a $\cap$-fuzzy probability measure on $\mathcal{A}$.

III. REALIZATIONS AND PLAUSIBILITY MEASURES

Let $(L,\otimes,\leq,c)$ be a $\otimes$-fuzzy σ-algebra. Each element μ of L is called an event, **1** and **0** are said to be the certain and the impossible event, respectively. A realization of L is a mapping $\omega : L \to \{0,1\}$ equipped with the following properties:

[R1] $\omega(\mathbf{0}) = 0$, $\omega(\mathbf{1}) = 1$; (boundary conditions)
[R2] $\omega(\mu \vee \upsilon) = \max(\omega(\mu),\omega(\upsilon))$. (ideal property)

The interpretation of a realization ω of L is as follows: an event $\mu \in L$ has occured with respect to ω (or is realized by the outcome ω) if and only if $\omega(\mu) = 1$.

3.1 Example. Let L be a σ-complete ortholattice (see example 1.1) and ω a realization of L. If an event μ does not occur with respect to ω then we infer from [R1] and [R2] that its complement $c(\mu)$ occurs.

The set I(L) of all realizations of L always will be equipped with the topology $\mathcal{T}_I$ generated by the following sub-base:

$$\mathcal{S} = \{\{\omega \in I(L) \mid \omega(\mu) = 0\} \mid \mu \in L\} \cup \{\{\omega \in I(L) \mid \omega(\mu) = 1\} \mid \mu \in L\}.$$

3.2 Lemma. $(I(L),\mathcal{T}_I)$ is a totally disconnected, compact space.

Proof. We choose subsets L_1 and L_2 of L such that

$$\mathcal{U} = \{\{\omega \mid \omega(\mu) = 0\} \mid \mu \in L_1\} \cup \{\{\omega \mid \omega(\mu) = 1\} \mid \mu \in L_2\}$$

is a covering of I(L). Without loss of generality we may assume $\mathbf{0} \notin L_1$ and $\mathbf{1} \notin L_2$; therefore L_2 is non-empty and we can consider the ideal I_o generated by L_2. Supposing $\{\omega \mid \omega(\mu) = 0\} \cap \{\omega \mid \omega(\upsilon) = 1\} \neq \phi$ for all $(\mu,\upsilon) \in I_o \times L_1$ we conclude that for all $U \in \mathcal{U}$ the realization $1_{CI_o} \in I(L)$ is not an element of U, hence $\mathcal{U}$ cannot be a covering of I(L). Therefore, by virtue of [R2], there exists a finite subcovering $\mathcal{V}$ of $\mathcal{U}$. Now the assertion follows from Alexander's sub-base theorem.

The next result establishes a canonical relationship between plausibility measures on L and a class of Borel probability measures on I(L) (see Höhle [7,8]).

3.3 Theorem. Let L be a ⊗-fuzzy σ-algebra and $m : L \to [0,1]$ be a mapping. Then the following assertions are equivalent:

(i) m is a plausibility measure on L;
(ii) there exists an unique $\mathcal{T}_I$-regular Borel probability measure υ_m on I(L) satisfying the subsequent conditions:

$$\forall\ \mu \in L\ :\ \upsilon_m(\{\omega | \omega(\mu) = 1\}) = m(\mu)\ ; \tag{3.1}$$

$$\forall\ (\mu_n)_{n\in\mathbb{N}} \in L^{\mathbb{N}},\ (\mu_n)\uparrow :$$

$$\upsilon_m(\{\omega|\omega(\bigvee_{n\in\mathbb{N}} \mu_n)=1\} \cap \complement(\bigcup_{n\in\mathbb{N}} \{\omega|\omega(\mu_n)=1\})) = 0. \tag{3.2}$$

Proof. The implication (ii) ⇒ (i) is evident. In order to verify (i) ⇒ (ii) we proceed as follows: since $\mathcal{E} = \{\{\omega|\omega(\mu) = 0\} | \mu \in L\}$ is closed under finite intersections the uniqueness of υ_m is a consequence of (3.1) and (3.2). Now let $\mathcal{A}$ be the algebra generated by $\mathcal{E}$; since $\mathcal{E}$ fulfills

$$\{\omega|\omega(\mu_o) = 0\} = \bigcup_{i=1}^{n} \{\omega|\omega(\mu_i) = 0\} \iff \tag{3.3}$$
$$\iff\ [\forall\ i \in \{1,\dots,n\}:\ \mu_o \leq \mu_i \text{ and } \exists\ i_o \in \{1,\dots,n\}:\ \mu_o \leq \mu_{i_o}]$$

we conclude from [M1], [M2] and [PL] that there exists a finitely additive probability measure π_m on $\mathcal{A}$ satisfying (3.1). Taking into account that each element of $\mathcal{A}$ is a $\mathcal{T}_I$-compact subset of I(L) (see lemma 3.1) we are able to extend π_m to a σ-additive probability measure $\overline{\pi_m}$ on the σ-algebra $\sigma(\mathcal{A})$ generated by $\mathcal{A}$. Since $\sigma(\mathcal{A})$ coincides with the σ-algebra of Baire subsets of I(L), $\overline{\pi_m}$ can be extended to a $\mathcal{T}_I$-regular Borel probability measure υ_m on I(L). By virtue of [M3] υ_m also fulfills (3.2).

The $\mathcal{T}_I$-regular Borel probability measure υ_m on I(L) existing because of the preceding theorem will be called the <u>canonical</u>

probability measure associated with m.

As an immediate consequence we obtain a result first proved by Shafer [23].

3.4 Corollary. Let L be a $\otimes$-fuzzy σ-algebra and $m : L \longrightarrow [0,1]$ be a mapping. Then the following assertions are equivalent:

(i) m is a plausibility measure on L;

(ii) there exists a probability σ-algebra $(\mathbb{B},P)$ (see Kappos [11]) and a $\vee$-homomorphism $h : L \longrightarrow \mathbb{B}$ satisfying the subsequent conditions:

$$h(\mathbf{0}) = \mathbf{0}, \; h(\mathbf{1}) = \mathbf{1} \; ; \tag{3.4}$$

$$\forall \, (\mu_n)_{n\in\mathbb{N}} \in L^{\mathbb{N}}, \; (\mu_n)\uparrow \; : \; h(\bigvee_{n\in\mathbb{N}} \mu_n) = \bigvee_{n\in\mathbb{N}} h(\mu_n); \tag{3.5}$$

$$\forall \, \mu \in L: \; m(\mu) = P(h(\mu)). \tag{3.6}$$

3.5 Corollary. Let $(L,\otimes,\leq,c)$ be a $\otimes$-fuzzy σ-algebra and M be a regular σ-complete sublattice of $(L,\leq)$ such that $(M,\otimes,\leq,c)$ is also a $\otimes$-fuzzy σ-algebra. Then each plausibility measure m on M can be extended to a plausibility measure $\overline{m}$ on L.

Proof. Let $(\mathbb{B}, P)$ be a probability σ-algebra and $h : M \longrightarrow \mathbb{B}$ be a $\vee$-homomorphism satisfying (3.4),(3.5) and (3.6). Since $\mathbb{B}$ is a weakly σ-distributive complete Boolean algebra (see Kappos [11]) h can be extended to a σ-smooth $\vee$-homomorphism $\overline{h} : L \longrightarrow \mathbb{B}$ (i.e. a $\vee$-homomorphism $\overline{h}$ fulfilling $\overline{h}(\bigvee_{n\in\mathbb{N}} \mu_n) = \bigvee_{n\in\mathbb{N}} \overline{h}(\mu_n)$ for all $(\mu_n)_{n\in\mathbb{N}} \in L^{\mathbb{N}}$) putting

$$\overline{h}(\mu) = \bigwedge \{h(\upsilon) \,|\, \upsilon \in M, \; \mu \leq \upsilon\}$$

for all $\mu \in L$. By virtue of corollary 3.4 $m^* = P \circ \overline{h}$ is a plausibility measure on L which by construction is an extension of m.

Finally we also can characterize $\otimes$-fuzzy probability measures m on L by a special class of Borel measures on I(L):

3.6 Corollary. Let $(L,\otimes,\leq,c)$ be a distributive fuzzy σ-algebra, m a plausibility measure on L and υ_m the canonical probability measure on I(L) associated with m. Then the following assertions are equivalent:

(i) m is a $\wedge$-fuzzy probability measure;
(ii) $\upsilon_m(\{\omega \mid \forall\, \mu,\upsilon \in L: \omega(\mu \wedge \upsilon) = \min(\omega(\mu),\omega(\upsilon))\}) = 1$.

IV. CLASSIFICATION OF FUZZY PROBABILITY MEASURES

Let $(X,\mathcal{A})$ be a measurable space; $\mathcal{B}$ shall denote the σ-algebra of Borel subsets of $[0,1[$. In what follows we shall characterize and classify T-fuzzy probability measures on $L = \zeta(\mathcal{A})$, the fuzzy σ-algebra generated by $\mathcal{A}$. A main tool in doing so will be the concept of Markov kernels. A Markov-kernel is a mapping $K : X \times \mathcal{B} \longrightarrow [0,1]$ being a measurable function in its first and a probability measure in its second component (see Bauer [2]).

4.1 Theorem. A mapping m is a $\wedge$-fuzzy probability measure on $L = \zeta(\mathcal{A})$ if and only if there exists a probability measure P on $\mathcal{A}$ and a Markov kernel K such that for all $\mu \in L$ we have

$$m(\mu) = \int_X K(x,[0,\mu(x)[)\,dP(x). \tag{4.1}$$

Proof. Obviously each mapping $m : L \longrightarrow [0,1]$ defined via (4.1) is a $\wedge$-fuzzy probability measure on L. Conversely, if m is a $\wedge$-fuzzy probability measure on L then $P(A) = m(1_A)$ for all $A \in \mathcal{A}$ defines a probability measure on $\mathcal{A}$; analogously for each $\alpha \in]0,1[$ $P_\alpha(A) = m(\alpha.1_A)$, $A \in \mathcal{A}$, yields a finite measure on $\mathcal{A}$ being absolutely continuous with respect to P. Therefore using the Radon-Nikodym Theorem we conclude that there is a family $(f_\alpha)_{\alpha\in\mathbb{Q}\cap]0,1[}$ of measurable functions $f_\alpha : X \longrightarrow [0,1]$ such that

$$P_\alpha(A) = \int_A f_\alpha \, dP$$

holds true for all $\alpha \in \mathbb{Q} \cap]0,1[$ and for all $A \in \mathcal{A}$. Now for each $\alpha \in [0,1[$ we consider the measurable function $g_\alpha = \sup\{f_\beta \mid \beta \in \mathbb{Q} \cap [0,\alpha[\}$; by the Lebesgue monotone convergence theorem and [M3] we get again

$$P_\alpha(A) = \int_A g_\alpha \, dP$$

for all $A \in \mathcal{A}$. Then for each $x \in X$ the mapping

$$h_x : [0,1[\longrightarrow [0,1]$$
$$\alpha \longrightarrow g_\alpha(x)$$

is a probability distribution function on $[0,1[$ inducing a probability measure Q_x on $\mathcal{B}$. The function

$$\begin{aligned} K : X\times\mathcal{B} &\longrightarrow [0,1] \\ (x,B) &\longrightarrow Q_x(B) \end{aligned} .$$

by definition is a probability measure in its second component; the measurability in the first component can be seen using a Dynkin system argument (see Bauer [2]). Hence K is a Markov kernel; the validity of (4.1) can then be established in the usual way proving it first for simple functions.

From this proof it follows immediately that the probability measure P is unique and that the Markov kernel K is P-almost surely determined. Therefore it makes sense to call the pair (P,K) the <u>Markov kernel representation</u> of the Λ-fuzzy probability measure m.

As a consequence of corollary 2.2(ii) each $\otimes$-fuzzy probability measure m on $L = \zeta(\mathcal{A})$ has a Markov kernel representation (P,K).

For the special family of T_s-fuzzy probability measures m, $s \in]0,+\infty]$, we can prove that the Markov-kernel representation is of a very special type.

<u>4.2 Theorem</u>. Let s be an element of $]0,+\infty]$, $L = \zeta(\mathcal{A})$ and m a T_s-fuzzy probability measure on L. Then there exists an unique probability P on $\mathcal{A}$ and a P-almost surely uniquely determined measurable function $f : X \longrightarrow [0,1]$ such that for all $\mu \in L$ we have

$$m(\mu) = \int_{\{\mu>0\}} [f + (1-f).\mu]\, dP. \tag{4.2}$$

Proof. Let (P,K) be the Markov kernel representation of m. Now we choose numbers $\alpha,\beta \in [0,1]$ such that $\alpha+\beta \leqq 1$ and an $A \in \mathcal{A}$ and define monotone sequences $(\mu_n)_{n\in\mathbb{N}}$, $(\upsilon_n)_{n\in\mathbb{N}}$ in L inductively by

$$\begin{aligned} \mu_1 &= \alpha.1_A , & \upsilon_1 &= \beta.1_A , \\ \mu_{n+1} &= S_s(\mu_n,\upsilon_n), & \upsilon_{n+1} &= T_s(\mu_n,\upsilon_n). \qquad (n\in\mathbb{N}) \end{aligned}$$

Using similar arguments as in the proof of theorem 1.3 we obtain

$$\sup_{n\in\mathbb{N}} \mu_n = (\alpha+\beta).1_A ,$$

$$\inf_{n\in\mathbb{N}} \upsilon_n = 1_\phi .$$

This subsequently leads to a Gaussian functional equation

(see Aczel [1]) for the Markov-kernel K whose unique solution is given by

$$K(x,[0,\alpha[) = f(x) + (1-f(x)).\alpha$$

for all $x \in X$ and all $\alpha \in [0,1]$, where $f : X \to [0,1]$ is some P-almost surely uniquely determined measurable function.

It should be noted that because of theorem 1.3 the assumption $L = \zeta(A)$ is always fulfilled; it has been used only to facilitate the statement of the preceding theorem.

In the case of $s \in]0,+\infty[$ the converse of theorem 4.2 is true, too:

<u>4.3 Proposition</u>. Let (X,A,P) be a probability space and $f : X \to [0,1]$ be a measurable function. Then for each $s \in]0,+\infty[$ the mapping m defined via (4.2) is a T_s-fuzzy probability measure on $L = \zeta(A)$.

Proof. See Klement [15] (theorem 4.1).

<u>4.4 Corollary</u>. Let s,t be elements of $]0,+\infty[$, $L = \zeta(A)$ and $m : L \to [0,1]$ a mapping. Then the following assertions are equivalent:

(i) m is a T_s-fuzzy probability measure on L;
(ii) m is a T_t-fuzzy probability measure on L.

In the case of T_∞-fuzzy probability measures, however, we are led back to the original definition of Zadeh [25]; the following result is also interesting from an axiomatic point of view: the fuzzy probability measures m_P as given by (2.1) are the only ones fulfilling additional properties such as σ-continuity or compatibility with the complement.

<u>4.5 Theorem</u>. Let s be an element of $]0,+\infty[$, $L = \zeta(A)$ and $m : L \to [0,1]$ a T_s-fuzzy probability measure on L. Then the following assertions are equivalent:

(i) $\forall\, (\mu_n)_{n\in\mathbb{N}} \in L^{\mathbb{N}}, \quad (\mu_n)\downarrow \;:\; m(\bigwedge_{n\in\mathbb{N}} \mu_n) = \inf_{n\in\mathbb{N}} m(\mu_n)$;
(ii) $\forall\, \mu \in L : \; m(1-\mu) = 1 - m(\mu)$;
(iii) there exists a probability measure P on A which induces m (see example 2.3(1));
(iv) m is a T_∞-fuzzy probability measure on L.

Proof. See Klement et al. [17] (theorem 2), Klement [15] (proposition 4.2).

In order to shed some light onto the relationship between the Markov kernel in the representation of a $\wedge$-fuzzy probability measure m on $\zeta(\mathcal{A})$ and the realizations of L we first observe that the mapping $i : X\times[0,1] \to I(\zeta(\mathcal{A}))$ defined by

$$(i(\alpha,x))(\mu) = 1 \quad \text{if } \mu(x) > \alpha,$$
$$= 0 \quad \text{if } \mu(x) \leq \alpha$$

for each $\mu \in \zeta(\mathcal{A})$ is Baire measurable with respect to the product σ-algebra $\mathcal{A} \times \mathcal{B}$ on $X \times [0,1]$. Furthermore, for each sequence $(\mu_n)_{n\in\mathbb{N}}$ of simple functions $\mu_n = \sum_{i=1}^{j_n} \alpha_i^{(n)} \cdot 1_{A_i^{(n)}}$ with $\alpha_1^{(n)},\ldots,\alpha_{j_n}^{(n)} \in [0,1]$, $A_1^{(n)},\ldots,A_{j_n}^{(n)}$ pairwise disjoint and $\bigcup_{i=1}^{j_n} A_i^{(n)} = X$ the following relation holds:

$$\{(\alpha,x) \in X\times[0,1] \mid \sup_{n\in\mathbb{N}} \mu_n(x) > \alpha\} =$$
$$= \bigcup_{n\in\mathbb{N}} [\bigcup_{i=1}^{j_n} (A_i^{(n)} \times [0,\alpha_i^{(n)}[)]. \tag{4.3}$$

If P is a probability measure on $\mathcal{A}$ and K a Markov kernel then a probability measure Q on $\mathcal{A}\times\mathcal{B}$ is induced via (see Klement and Schwyhla [18])

$$Q(H) = \int_X K(x,H_x)\, dP(x) \tag{4.4}$$

for all $H \in \mathcal{A}\times\mathcal{B}$, where H_x denotes the x-section of H, that means $H_x = \{\alpha \in [0,1[\mid (x,\alpha) \in H\}$.

<u>4.6 Proposition</u>. Let m be a $\wedge$-fuzzy probability measure on $\zeta(\mathcal{A})$, (P,K) its Markov kernel representation, υ_m the canonical probability measure associated with m and Q the probability measure on $\mathcal{A}\times\mathcal{B}$ induced by P and K via (4.4). Then the mapping $i : X\times[0,1[\to I(\zeta(\mathcal{A}))$ is measure preserving, i.e. we have $Q(i^{-1}(B)) = \upsilon_m(B)$ for all Baire subsets B of $I(\zeta(\mathcal{A}))$.

Proof. Let $(\mu_n)_{n\in\mathbb{N}}$ be an increasing sequence of simple functions.

Then using (4.3) we obtain:

$$\upsilon_m(\{\omega \in I(\zeta(A)) \mid \omega(\sup_{n\in\mathbb{N}} \mu_n) = 1\}) = m(\sup_{n\in\mathbb{N}} \mu_n) = \sup_{n\in\mathbb{N}} m(\mu_n) =$$

$$= \sup_{n\in\mathbb{N}} \left(\sum_{i=1}^{j_n} \int_{A_i^{(n)}} K(x, [0,\alpha_i^{(n)}[\,)\, dP(x)\right)$$

$$= \sup_{n\in\mathbb{N}} Q\left(\bigcup_{i=1}^{j_n} (A_i^{(n)} \times [0,\alpha_i^{(n)}[)\right) = Q(i^{-1}(\{\omega \mid \omega(\sup_{n\in\mathbb{N}} \mu_n) = 1\})).$$

This proposition shows that the Markov kernel representation of a Λ-fuzzy probability measure m on $\zeta(A)$ can be identified with the canonical probability measure υ_m on the set of all realizations of $\zeta(A)$. In this sense the Markov kernel representation illustrates clearly the special structure of the canonical probability measure υ_m associated with the Λ-fuzzy probability measure m on $\zeta(A)$.

V. ENTROPIES WITH RESPECT TO PLAUSIBILITY MEASURES

Let m be a plausibility measure on the fuzzy σ-algebra $(L,\wedge,\leq,c)$. We now consider mappings e_m, s_m^1, s_m^2 from I(L), the set of all realizations of L, into $[0,+\infty]$ defined by

$$e_m(\omega) = \sup\{(-\ln(m(\mu \vee c(\mu)) - m(\mu))) \cdot (1-\omega(\mu)) \cdot \omega(c(\mu)) \mid \mu \in L\}, \tag{5.1}$$

$$s_m^1(\omega) = \sup\{(-\ln(m(\mu) + m(c(\mu)) - m(\mu \vee c(\mu)))) \cdot \omega(\mu) \cdot \omega(c(\mu)) \mid \mu \in L\}, \tag{5.2}$$

$$s_m^2(\omega) = \sup\{(-\ln(1 - m(\mu \vee c(\mu)))) \cdot (1-\omega(\mu)) \cdot (1-\omega(c(\mu))) \mid \mu \in L\} \tag{5.3}$$

for all $\omega \in I(L)$.

An interpretation of the (possibly infinite) values of these mappings can be given as follows: an event μ is <u>discernible</u> from $c(\mu)$ with respect to a realization ω if, and only if, $\omega(\mu) \neq \omega(c(\mu))$; μ is <u>indiscernible</u> from $c(\mu)$ with respect to ω if, and only if, $\omega(\mu) = \omega(c(\mu))$; in this context $e_m(\omega)$ is the maximal information that ω distinguishes between μ and $c(\mu)$, while $s_m^1(\omega)$ and $s_m^2(\omega)$ give the maximal information that μ and $c(\mu)$ are indistinguishable with respect to ω. Since e_m, s_m^1 and s_m^2 are Borel measurable functions it is possible to define the <u>entropy of discernibleness</u> $E_m(L)$ and the <u>entropies of indiscernibleness</u> $S_m^1(L)$ and $S_m^2(L)$ of the <u>first</u> and <u>second kind</u>, respectively, as the mean value of the information given by e_m, s_m^1 and s_m^2, respectively, that means

$$E_m(L) = \int_{I(L)} e_m \, d\upsilon_m, \tag{5.4}$$

$$S_m^1(L) = \int_{I(L)} s_m^1 \, d\upsilon_m, \tag{5.5}$$

$$S_m^2(L) = \int_{I(L)} s_m^2 \, d\upsilon_m, \tag{5.6}$$

where υ_m is the canonical probability measure associated with m.

Now we calculate these entropies for several special plausibility measures (Höhle [9]).

5.1 Proposition. Let m be a probability measure on $\mathcal{P}(\mathbb{N})$, the power set of $\mathbb{N}$. Then the entropies of discernibleness and indiscernibleness are given by

$$E_m(\mathcal{P}(\mathbb{N})) = \sum_{n=1}^{\infty} -m(\{n\}).\ln(m(\{n\})), \tag{5.7}$$

$$S_m^1(\mathcal{P}(\mathbb{N})) = S_m^2(\mathcal{P}(\mathbb{N})) = 0. \tag{5.8}$$

In particular the entropy of discernibleness coincides with Shannon's entropy.

Proof. Each $n \in \mathbb{N}$ induces a realization $\omega_n \in I(\mathcal{P}(\mathbb{N}))$ by

$$\omega_n(A) = 1 \quad \text{if } n \in A,$$
$$\qquad = 0 \quad \text{if } n \notin A$$

for all $A \subset \mathbb{N}$. Obviously $\Phi : \mathbb{N} \rightarrow I(\mathcal{P}(\mathbb{N}))$ given by $\Phi(n) = \omega_n$ for all $n \in \mathbb{N}$ is measure preserving; therefore (5.7) follows from

$$E_m(\mathcal{P}(\mathbb{N})) = \int_{\mathbb{N}} e_m \circ \Phi \, dm.$$

The proof of (5.8) is obvious.

5.2 Proposition. Let X be a non-empty set, f be a normal fuzzy subset of X and m_f the $\cap$-possibility measure on $\mathcal{P}(X)$ induced by f (see example 2.3(2)). Then the entropies $E_{m_f}(\mathcal{P}(X))$, $S_{m_f}^1(\mathcal{P}(X))$ and $S_{m_f}^2(\mathcal{P}(X))$ are given by

$$E_{m_f}(\mathcal{P}(X)) = \int_0^1 -\ln(1 - \sup f(f^{-1}([0,\alpha[)))d\alpha, \tag{5.9}$$

$$S^1_{m_f}(P(X)) = \int_0^1 -\ln(\inf f(f^{-1}([\alpha,1])))d\alpha, \qquad (5.10)$$

$$S^2_{m_f}(P(X)) = 0 \qquad (5.11)$$

Proof. Since $P(X)$ is a Boolean algebra the validity of (5.11) is an immediate consequence of the definition of $S^2_{m_f}(P(X))$ (see example 3.1). In order to verify (5.9) and (5.10) we proceed as follows: each $\alpha \in [0,1]$ induces a realization $\omega_\alpha \in I(P(X))$ given by

$$\omega_\alpha(A) = 1 \quad \text{if} \quad A \cap \{f \geqq \alpha\} \neq \phi,$$
$$\quad = 0 \quad \text{if} \quad A \cap \{f \geqq \alpha\} = \phi$$

for all A X; then we define a function $\Phi : [0,1] \longrightarrow I(P(X))$ by $\Phi(\alpha) = \omega_\alpha$ for all $\alpha \in [0,1]$. If we equip [0,1] with the Lebesgue measure then Φ is measure preserving. Furthermore the following relations are valid for all $\alpha \in [0,1]$:

$$\sup\{-\ln(1 - m_f(A) \mid A \cap \{f \geqq \alpha\} = \phi\} = -\ln(1 - \sup f(f^{-1}([0,\alpha[))),$$

$$\sup\{-\ln(m_f(A) + m_f(\complement A) - 1) \mid A \cap \{f \geqq \alpha\} \neq \phi,\ \complement A \cap \{f \geqq \alpha\} \neq \phi\} \leqq$$

$$\leqq \ln(\inf f(f^{-1}([\alpha,1]))),$$

$$-\ln(\inf f(f^{-1}([\alpha,1]))) =$$

$$= \sup\{-\ln(m_f(\{x\}) + m_f(\complement\{x\}) - 1) \mid \alpha \leqq f(x) < 1\} \leqq$$

$$\leqq \sup\{-\ln(m_f(A) + m_f(\complement A) - 1 \mid A \cap \{f \geqq \alpha\} = \phi,\ \complement A \cap \{f \geqq \alpha\} = \phi\}.$$

Consequently we get

$$\int_0^1 e_{m_f} \circ \Phi(\alpha)\, d\alpha = \int_0^1 -\ln(1 - \sup f(f^{-1}([0,\alpha[)))\, d\alpha,$$

$$\int_0^1 s^1_{m_f} \circ \Phi(\alpha)\, d\alpha = \int_0^1 -\ln(\inf f(f^{-1}([\alpha,1[)))\, d\alpha,$$

thus verifying (5.9) and (5.10).

<u>5.3 Proposition</u>. Let (X,A,P) be a probability space and let m_P be the $\wedge$-fuzzy probability measure on $\zeta(A)$ induced by P (see example 2.3(1)). Then the entropies $E_{m_P}(\zeta(A))$, $S^1_{m_P}(\zeta(A))$ and $S^2_{m_P}(\zeta(A))$ are given by

$$E_{m_P}(\zeta(A)) = E_P(A) + 1 , \tag{5.12}$$

$$S^1_{m_P}(\zeta(A)) = S^2_{m_P}(\zeta(A)) = \frac{1}{2}.(E_P(A) + 1 + \ln 2). \tag{5.13}$$

Proof. Let $H(A)$ be the set of all Boolean homomorphisms from A onto $\{0,1\}$, i.e.

$$H(A) = \{\varphi \in I(A) \mid \varphi(A \cap B) = \min(\varphi(A),\varphi(B))\}.$$

By virtue of theorem 3.3 and corollary 3.6 the canonical probability measure υ_P on $I(A)$ fulfills (3.1) and lives on $H(A)$, i.e. $\upsilon_P(H(A)) = 1$. Now each pair $(\varphi,\alpha) \in H(A)\times[0,1]$ induces a realization $\omega_{(\varphi,\alpha)} \in I(\zeta(A))$ by

$$\omega_{(\varphi,\alpha)}(\mu) = \varphi(\mu^{-1}([\alpha,1]))$$

for all $\mu \in \zeta(A)$. If we equip $H(A)\times[0,1]$ with $\upsilon_P\times\lambda$, the product of the canonical probability measure υ_P associated with P and the Lebesgue measure λ, then the mapping $\Phi : H(A)\times[0,1] \longrightarrow I(\zeta(A))$ defined by $\Phi(\varphi,\alpha) = \omega_{(\varphi,\alpha)}$ for all $(\varphi,\alpha) \in H(A)\times[0,1]$ is measure preserving. Referring to standard methods in measure theory it is not difficult to do the following calculations:

$$E_{m_P}(\zeta(A)) =$$

$$= \int_{H(A)\times[\frac{1}{2},1]} e_{m_P} \circ \Phi \, d(\upsilon_P\times\lambda) + \int_{H(A)\times[0,\frac{1}{2}]} e_{m_P} \circ \Phi \, d(\upsilon_P\times\lambda) =$$

$$= \int_{H(A)\times[\frac{1}{2},1]} \sup\{-\ln((2\alpha-1).P(A)).\varphi(A) \mid A \in A\} \, d(\upsilon_P\times\lambda)(\varphi,\alpha) +$$

$$+ \int_{H(A)\times[0,\frac{1}{2}]} \sup\{-\ln((1-2\alpha).P(A)).\varphi(A) \mid A \in A\} \, d(\upsilon_P\times\lambda)(\varphi,\alpha) =$$

$$= E_P(A) + 1 ;$$

$$S^2_{m_P}(\zeta(A)) = S^1_{m_P}(\zeta(A)) = \int_{H(A)\times[0,\frac{1}{2}]} s^1_{m_P} \circ \Phi \, d(\upsilon_P\times\lambda) =$$

$$= \int_{H(A)\times[0,\frac{1}{2}]} \sup\{-\ln(\alpha.P(A)).\varphi(A) \mid A \in A\} \, d(\upsilon_P\times\lambda)(\varphi,\alpha) =$$

$$= \frac{1}{2}.(E_P(A) - \ln\frac{1}{2} + 1).$$

Therewith the assertions are verified.

Let T be a measurable t-norm and m be a T-fuzzy probability measure on a T-fuzzy σ-algebra L. If m is not a T_∞-fuzzy probability measure then theorem 4.5 shows that the entropies of discernibleness of the first and of the second kind are different from each other and different from zero.

VI. POSSIBLE FUZZIFICATIONS OF THE THEORY

What we have done up to this point is classical measure theory on lattices with different kinds of additivity. Now it might be interesting to fuzzify the theory, that means to replace crisp sets by fuzzy ones.

The fuzzification of the events has been suggested by Zadeh [25] who considered a generated fuzzy σ-algebra and called its elements _fuzzy events_. However, since each distributive fuzzy σ-algebra may be embedded into a Boolean algebra (see Birkhoff [3]) we again come back to the classical case.

A more promising way should be to deal with _fuzzy realizations_ which map into [0,1] instead of {0,1} (Höhle [8]).

A third idea is to fuzzify the range of the measures considering _fuzzy real numbers_ which could include the treatment of things like linguistic information and errors of measurement into the theory. In some sense the theory developed by Klement [16] also can be regarded as a _stochastic possibility theory_.

REFERENCES

1. Aczel, J.: 1969, _Lectures on Functional Equations and Their Applications_, Academic Press, New York.
2. Bauer, H.: 1981, _Probability Theory and Elements of Measure Theory_, Academic Press, London.
3. Birkhoff, G.: 1967, _Lattice Theory_, Amer. Math. Soc. Colloq. Publ., Vol. XXXV (Third Edition).
4. Frank, M.J.: 1979, 'On the simultaneous associativity of F(x,y) and x+y-F(x,y)', _Aequationes Math._ 19, 194-226.
5. Fuchs, L.: 1966, _Teilweise geordnete algebraische Strukturen_, Vandenhoeck & Ruprecht, Göttingen.
6. Höhle, U.: 1976, 'Maße aus unscharfen Mengen', _Z. Wahrscheinlichkeitstheorie verw. Gebiete_ 36, 179-188.

7. Höhle, U.: 1982, 'A mathematical.theory of uncertainty', in R.R. Yager (ed.), Fuzzy Set and Possibility Theory. Recent Developments, Pergamon Press, New York, pp. 344-355.
8. Höhle, U.: 1982, 'Fuzzy plausibility measures', in E.P. Klement (ed.), Proceedings of the Third International Seminar on Fuzzy Set Theory, Linz, 1981, Johannes Kepler Universität, Linz, pp. 7-30.
9. Höhle U.: 1982, 'Entropy with respect to plausibility measures', in Proceedings, The Twelfth International Symposium on Multiple-valued Logic, Paris, 1982, IEEE, New York, pp. 167-169.
10. Jakobs, K.: 1978, Measure and Integral, Academic Press, New York.
11. Kappos, D.A.: 1969, Probability algebras and stochastic spaces, Academic Press, New York.
12. Klement, E.P.: 1980, 'Fuzzy σ-algebras and fuzzy measurable functions', Fuzzy Sets and Systems 4, 83-93.
13. Klement, E.P.: 1980, 'Characterization of finite fuzzy measures using Markoff-kernels', J. Math. Anal. Appl. 75, 330-339.
14. Klement, E.P.: 1982, 'Construction of fuzzy σ-algebras using triangular norms', J. Math. Anal. Appl. 85, 543-565.
15. Klement, E.P.: 1982, 'Characterization of fuzzy measures constructed by means of triangular norms', J. Math. Anal. Appl. 86, 345-358.
16. Klement, E.P.: 'Fuzzy measures assuming their values in the set of fuzzy numbers', J. Math. Anal. Appl., in press.
17. Klement, E.P., Lowen, R. and Schwyhla, W.: 1981, 'Fuzzy probability measures', Fuzzy Sets and Systems 5, 21-30.
18. Klement, E.P. and Schwyhla, W.: 1982, 'Correspondence between fuzzy measures and classical measures', Fuzzy Sets and Systems 7, 57-70.
19. Łos, J.: 1955, 'On the axiomatic treatment of probability' Colloq. Math. 3, 125-137.
20. Łos, J.: 1960, 'O ciatach zdarzeń i ich definicji w aksjomatycznej teorii prawdopodobieństwa', Studia Logica 9, 95-132. ('Fields of events and their definitions in the axiomatic treatment of probability theory', in Selected Translations of Mathematical Statistics and Probability, Vol. 7, Inst. of Math. Stat., Amer. Math. Soc., Providence, 1968, pp. 17-39)

21. Schweizer, B. and Sklar, A.: 1963, 'Associative functions and abstract semigroups', Publ. Math. Debrecen 10, 69-81.
22. Shafer, G.: 1976, A Mathematical Theory of Evidence, Princeton University Press, Princeton.
23. Shafer, G.: 1979, 'Allocations of probability', Annals of Prob. 7, 827-839.
24. Zadeh, L.A.: 1965, 'Fuzzy Sets', Inform. Control 8, 338-353.
25. Zadeh, L.A.: 1968, 'Probability measures of fuzzy events', J. Math. Anal. Appl. 23, 421-427.
26. Zadeh, L.A.: 1978, 'Fuzzy Sets as a basis for a theory of possibility', Fuzzy Sets and Systems 1, 3-28.

Michael Katz

CONTROLLED-ERROR THEORIES OF PROXIMITY AND DOMINANCE

1. INTRODUCTION

This paper, although self-contained and self-explanatory, can be considered complementary to the paper I presented in Las Palmas [1982a] at the Second World Conference on Mathematics at the Service of Man. The two papers deal with theories of proximity and dominance (in the context of social and behavioral inquiry) and they are complementary in the following senses:
(a) In [1982a] we were interested in representing ("measuring") proximity and dominance by intervals on the real line. The theories served as tools guaranteeing the existence of the required representations. In this paper the stress is on the theories themselves - their logic and their meaning. However, we ignore here theories requiring an informal mathematical language (which were quoted in [1982a] from works of other researchers such as Scott and Suppes [1958], Roberts [1969] and Krantz et al. [1971]) and limit ourselves to (parts of) those which can be axiomatized in a formal (two-valued or multi-valued) language. These theories can also be found in our [1981a] and [1981b].
(b) While our theories in [1982a] were either "error-free" or "error-bound", and dealt with proximity or with dominance, here we introduce "controlled-error" theories which tie the two predicates together. These theories (to be found also in our [1981a] and [1981b]) involve "natural deductions", unlike the theories in [1982a] which only contain "axioms". (All the terms in inverted commas in this paragraph will be explained below).

In the following section we present certain parts of the formal theories studied in our [1982a]. We move to controlled-error theories, axiomatized in the "logic of approximation", in Section 3. Some results of these theories, in particular results concerning the notion of substitutability, are considered in Section 4. Finally, in Section 5, we discuss several philosophical implications of the logical systems behind our theories. From the logical-philosophical

H. J. Skala, S. Termini, and E. Trillas (eds.), Aspects of Vagueness, 51–74.

point of view the scope of this paper is much broader, and the handling of error and approximation much subtler, than what we have in [1982a].

2. AXIOMATIC THEORIES

We are concerned here with formal theories in the behavioral sciences, and according to the influential work of Coombs [1964] the two basic predicates of such theories should be those of proximity and dominance, since virtually any type of behavioral data can be reduced to these two predicates. That proximity (or similarity) relations are indeed ubiquitous in behavioral theory and research can be learnt from the works of, e.g., Gregson [1975] and Shepard [1974]. The importance of dominance (or preference) is stressed, e.g., in the classical book of Luce and Raiffa [1957] on games and decisions, where a central role is also played by the indifference relation (a close relative of proximity).

Thus we shall assume now that we have a first-order language L, with variables x,y,z (with or without subscripts), whose atomic formulae are 2-place formulae of either the form $x \sim y$ (asserting that the objects to which x and y refer are proximate) or the form $x \precsim y$ (asserting that the object to which x refers dominates the one to which y refers). Compound formulae are built in the usual way, using the connectives $\neg, \wedge, \vee, \rightarrow$ (and the auxiliary symbols of comma and parentheses, if and when needed). Quantifiers may be added to the language (as in, e.g., our [1982b]), but this will not be necessary for our purposes here. (Only at the end of Section 4 below we shall briefly consider languages with quantifiers and with n-place atomic formulae for various n's). Arbitrary formulae of L will be denoted by the Greek letters φ, ψ, θ.

A structure $\underline{A}$ for L consists of a non-empty set A (the "domain" of $\underline{A}$) and two [0,1]-valued functions, $[\![\sim]\!]$ and $[\![\precsim]\!]$, on A^2 (the "interpretations" of $\sim$ and $\precsim$, respectively, in $\underline{A}$). If a and b are elements of A we write $[\![a \sim b]\!]$ for $[\![\sim]\!](a,b)$ and $[\![a \precsim b]\!]$ for $[\![\precsim]\!](a,b)$. If φ is a formula of L, X a set of variables of L containing all variables of φ, and $\bar{a}$ a function from X into the domain A of $\underline{A}$, then $\varphi\bar{a}$ is the "truth-value" of φ at $\bar{a}$. These truth-values in $\underline{A}$ are defined inductively as follows.

$$(x \sim y)\bar{a} = [\![\bar{a}(x) \sim \bar{a}(y)]\!] \qquad (1)$$

$$(x \precsim y)\bar{a} = [\![\bar{a}(x) \precsim \bar{a}(y)]\!] \tag{2}$$

$$(\neg\psi)\bar{a} = 1 - \psi\bar{a} \tag{3}$$

$$(\psi \wedge \theta)\bar{a} = \max\{\psi\bar{a}, \theta\bar{a}\} \tag{4}$$

$$(\psi \vee \theta)\bar{a} = \min\{\psi\bar{a}, \theta\bar{a}\} \tag{5}$$

$$(\psi \rightarrow \theta)\bar{a} = \max\{0, \theta\bar{a} - \psi\bar{a}\} \tag{6}$$

In our multi-valued logic, as described here, truth-values represent degrees of error. Thus 0 (no-error) stands for absolute truth and 1 (maximal error) for absolute falsity. Clauses (3)-(6) above are (reversed) extensions of the semantic rules of Lukasiewicz Logic (see, e.g., pp. 38-59 in Tarski [1956]), and they reduce to the well-known truth-tables of two-valued classical logic if the only admissible values are 0 and 1. The idea behind these clauses is that a conjunction maximizes the error while a disjunction minimizes it; that the error in an implication assertion is the degree to which the antecedent is truer than the consequent (or 0 if the consequent is truer than the antecedent); and that a negation assertion is just an implication assertion in which the consequent is absolute falsity. Similar ideas can be found in Giles [1974] and in Scott [1974].

These remarks also motivate our next definition. We write $\vdash\varphi$ to say that the formula φ of L is an "axiom", and we call the structure $\underline{A}$ for L a "model" of the axiom if φ is constantly fully true in $\underline{A}$, i.e., if for every function $\bar{a}$ as above (mapping the variables of φ into the domain A of $\underline{A}$) we have $\varphi\bar{a} = 0$. An axiomatic theory is a list of axioms, and a given structure for L is a model of a given axiomatic theory if it is a model of every axiom in the theory.

We can turn now to axiomatic theories of proximity and dominance, and we start with proximity. The theory P of proximity is the following list of axioms (where [RE] stands for reflexivity, [SY] for symmetry and [TR] for transitivity).

[RE] $\vdash x \sim x$

[SY] $\vdash x \sim y \rightarrow y \sim x$

[TR] $\vdash x \sim y \rightarrow (y \sim z \rightarrow x \sim z)$.

It is easy to see that in our logic of error the structure $\underline{A}$ for L is a model of this theory if and only if for every a,b,c in the domain A of $\underline{A}$ we have

$$[\underline{RE}] \quad [\![a \sim a]\!] = 0$$

$$[\underline{SY}] \quad [\![a \sim b]\!] = [\![b \sim a]\!]$$

$$[\underline{TR}] \quad [\![a \sim c]\!] \leqslant [\![a \sim b]\!] + [\![b \sim c]\!] \quad .$$

Thus $\sim$ is an equivalence relation in L but $[\![\sim]\!]$ is a pseudo-metric in $\underline{A}$. Our pseudo-metric reduces to the discrete pseudo-metric if we only allow the values 0 and 1, in which case we have a classical logic and we can replace [TR] by $\vdash (x \sim y \wedge y \sim z) \rightarrow x \sim z$. This last axiom translates in $\underline{A}$ to the ultrametric inequality, $\max([\![a \sim b]\!],[\![b \sim c]\!]) \geqslant [\![a \sim c]\!]$ (see Shepard [1974]), which holds for the discrete metric but not for an arbitrary metric. As Scott [1974] noted, we have here an interesting and intuitively meaningful illustration of the well-known fact that while the axioms $\vdash (\varphi \wedge \psi) \rightarrow \theta$ and $\vdash \varphi \rightarrow (\psi \rightarrow \theta)$ are equivalent in two-valued logic the former is stronger than the latter in multi-valued logic.

In our Las-Palmas paper [1982a] we supplemented the theory P with two more axioms in L, guaranteeing that for every model $\underline{A}$ of the extended theory there is a real-valued function f on the domain A of $\underline{A}$ such that for every a and b in A

$$[\![a \sim b]\!] = |f(a) - f(b)| \quad .$$

Here we shall not be interested in these axioms or in the measurement function f on A. Instead, we proceed to consider our theory of dominance, D, which is the following list of axioms (where [CO] stands for connectedness and [DT] for dominance-transitivity):

$$[CO] \quad \vdash x \gtrsim y \vee y \gtrsim x$$

$$[DT] \quad \vdash x \gtrsim y \rightarrow (y \gtrsim z \rightarrow x \gtrsim z).$$

Clearly, a structure $\underline{A}$ for L with domain A is a model of this theory if and only if for every a,b,c in A we have

$$[\underline{CO}] \quad \min\{[\![a \gtrsim b]\!],[\![b \gtrsim a]\!]\} = 0$$

$$[\underline{DT}] \quad [\![a \gtrsim c]\!] \leqslant [\![a \gtrsim b]\!] + [\![b \gtrsim c]\!].$$

Thus $\gtrsim$ is a weak ordering relation in L, while $[\![\gtrsim]\!]$ satisfies an obvious minimality condition and a triangle inequality in $\underline{A}$. The triangle inequality in this context tells us that the error in the assertion that a dominates c cannot exceed the combined error in the assertions that a dominates b and b dominates c.

This theory of dominance was supplemented in our [1982a] with an additional axiom to guarantee the existence of a measurement function f on the domain A of an appropriate model $\underline{A}$ such that for all a and b in A

$$[\![a \gtrsim b]\!] = \max\{0, f(b) - f(a)\}.$$

Once again, we shall not be interested here in this imbeddability in the real line. Instead, we conclude this section by noting that every model $\underline{A}$ of the theory D of dominance is also a model of the following axioms of dominance-reflexivity and reversed-transitivity ([DR] and [RT], respectively).

[DR] $\vdash x \gtrsim x$

[RT] $\vdash x \gtrsim y \rightarrow (z \gtrsim x \rightarrow z \gtrsim y).$

Much as in the case of classical weak-ordering, [DR] follows immediately from [CO] in $\underline{A}$ while [RT] is obviously equivalent to [DT] in $\underline{A}$.

3. DEDUCTIVE THEORIES

We now turn to deal with theories tying proximity and dominance together. These, unlike the axiomatic theories of the preceding section, are "deductive" theories. They involve "natural deductions", i.e., expressions of the form $\Gamma \vdash \varphi$, asserting that φ follows from Γ, where Γ is a (finite) set of formulae of L.

Another important distinction is this. The theories of the preceding section were interpreted in a structure $\underline{A}$ for L as either "error-free" (inducing, for instance, in the case of proximity, a discrete pseudo-metric on the domain A of $\underline{A}$) or "error-bound" (inducing in the case of proximity an arbitrary pseudo-metric on A). In the first case, only the values 0 ("true") and 1 ("false") are allowed. In the second, we may have any degree of error in the interval [0,1], and we are unable to control this error. The theories we shall have now, in contrast, are based on the assumption that it is within the power of the theorist, at least in principle, to control the error in the assertion φ by "playing" with the assumptions in the set Γ. This is why we use here the term "controlled-error theories".

Let us start again with the language, structures and multi-valued semantic rules of the preceding section. A structure $\underline{A}$ for L is a model of the deduction expression $\Gamma \vdash \varphi$, in what we called in, e.g., [1981c] and [1982c] "the logic of approximation" (LA), if for every ε we can find a δ such that for every appropriate $\bar{a}$

$$(\wedge\Gamma)\bar{a} < \delta \Rightarrow \varphi\bar{a} < \varepsilon.$$

Here, and in the sequel, ε's and δ's denote positive real-numbers, $\Rightarrow$ denotes implication in the meta-language, $\wedge\Gamma$ is the conjunction of all formulae in Γ, and by "appropiate $\bar{a}$" we mean a function from the set of variables of the formulae in $\Gamma \cup \{\varphi\}$ into the domain A of $\underline{A}$. When we write below an explicit deduction expression we shall usually simply list the elements of Γ on the left side of the symbol $\vdash$, omitting the set brackets.

A deduction expression, $\Gamma \vdash \varphi$, reduces to an axiom when Γ is empty. And then being a model of $\Gamma \vdash \varphi$, in the sense of this section, reduces to being a model of $\vdash \varphi$, in the sense of the preceding section, by the convention $\max \phi = 0$, where ϕ is the empty set. Thus a deductive theory may well be a mixed list of deductions and axioms (as the latter are nothing but deductions from the empty set).

This brings us to the theory PD, of proximity and dominance, which consists of the axioms of P, the axioms of D, and the following deductions (where [SR] stands for strong reflexivity, as it is obviously stronger in this theory than [DR], and [AS] stands for asymmetry):

[SR] $\quad x \sim y \vdash x \gtrsim y$

[AS] $\quad x \gtrsim y,\ y \gtrsim x \vdash x \sim y$.

A structure $\underline{A}$ for L with domain A is a model of these two deductions if for every ε there are δ_1 and δ_2 such that for each a and b in A

[$\underline{\text{SR}}$] $\quad [\![a \sim b]\!] < \delta_1 \Rightarrow [\![a \gtrsim b]\!] < \varepsilon$

[$\underline{\text{AS}}$] $\quad \max\{[\![a \gtrsim b]\!], [\![b \gtrsim a]\!]\} < \delta_2 \Rightarrow [\![a \sim b]\!] < \varepsilon$.

Thus, in the formal language L the theory PD (containing the theories P and D) makes proximity and dominance (and their inter-relations) completely analogous to equality and ordering (and their inter-relations). Yet in an actual model $\underline{A}$ of PD the properties relating equality and ordering are satisfied by proximity and dominance only in an appropriate manner. Some consequences of this idea will be discussed in the next section.

Here we present another deductive theory, this time relating to "strict" dominance, although we shall not study this theory separately in subsequent sections of this paper. We assume now that instead of atomic formulae of the form $x \succsim y$, L contains formulae of the form $x \succ y$ (as well as compound formulae, built up from these new formulae and/or from those of the form $x \sim y$). Similarly, instead of the function $[\![\succsim]\!]$ a structure $\underline{A}$ for L with domain A contains a function $[\![\succ]\!]$ (also from A^2 into $[0,1]$). We write $[\![a \succ b]\!]$ for $[\![\succ]\!](a,b)$ and $[\![\bar{a}(x) \succ \bar{a}(y)]\!]$ for $(x \succ y)\bar{a}$, where a and b are in A and $\bar{a}$ is as in Section 2 above.

The theory PSD of proximity and strict dominance, in the new language L, consists of the axioms of P together with the following list of deductions (where [ST] stands for strict transitivity, [IR] for irreflexivity and [SC] for strict connectedness):

$[ST]_1$ $\quad x \succ y \vdash (y \succ z \rightarrow x \succ z)$

$[ST]_2$ $\quad x \succ y \vdash (z \succ x \rightarrow z \succ y)$

[IR] $\quad x \sim y \vdash \neg(x \succ y)$

[SC] $\quad \neg(x \succ y), \neg(y \succ x) \vdash x \sim y.$

A structure $\underline{A}$ for the new L is a model of these four deduction expressions if for every ε we can find $\delta_1, \delta_2, \delta_3, \delta_4$, such that for all a,b,c in the domain A of $\underline{A}$

$[\underline{ST}]_1$ $\quad [\![a \succ b]\!] < \delta_1 \Rightarrow [\![a \succ c]\!] - [\![b \succ c]\!] < \varepsilon$

$[\underline{ST}]_2$ $\quad [\![a \succ b]\!] < \delta_2 \Rightarrow [\![c \succ b]\!] - [\![c \succ a]\!] < \varepsilon$

$[\underline{IR}]$ $\quad [\![a \sim b]\!] < \delta_3 \Rightarrow 1 - [\![a \succ b]\!] < \varepsilon$

$[\underline{SC}]$ $\quad \max\{1-[\![a \succ b]\!], 1-[\![b \succ a]\!]\} < \delta_4 \Rightarrow [\![a \sim b]\!] < \varepsilon.$

In the formal language L the theory PSD (containing the theory P) provides proximity and strict dominance with all the properties and inter-relations of equality and strict ordering. Yet, in a model of PSD some of these properties hold only up to an appropriate approximation.

Note that since we have transitivity here in the form of deduction expressions, rather than in the (stronger) form of axioms, we cannot derive $[ST]_2$ from $[ST]_1$ in the way we derived [RT] from [DT] in the (axiomatic) case of non-strict dominance (at the end of the preceding section). However, it is easy to see that from [IR] we can derive the following axiomatic form of irreflexivity, $[IR]_0$, and that then from $[IR]_0$ and $[ST]_1$ we can derive the anti-symmetry deduction [AN] (in every model of PSD):

$[IR]_0$ $\quad \vdash \neg(x \succ x)$

[AN] $\quad x \succ y \vdash \neg(y \succ x)$.

In the following section we consider some theorems about substitutability of proximates. It will be the only section of this paper to contain such mathematical "technicalities" as theorems and proofs. Up until now we presented "ideas": syntactic notions of axiomatic and deductive theories, corresponding semantic notions of error-bound and controlled-error theories, specific syntactic theories of proximity and dominance, and direct translations of these specific theories in appropriate structures. We shall return to a discussion of these and other, closely related, ideas in Section 5 below, the closing section of this paper.

4. SUBSTITUTABILITY

From the moment we introduced above our ε— and —δ interpretation of deductions it must have been clear that the notion of continuity will turn up somewhere in our discussion. Indeed, so it does, and in the very plausible context of substitutability.

To complete the analogy between proximity and dominance on the one hand and equality and ordering on the other hand, we consider now substitutability of proximates in dominance formulae. We return here to our original language L with atomic formulae of the forms $x \sim y$ and $x \succsim y$ (and, accordingly, to structures with the functions $[\![\sim]\!]$ and $[\![\succsim]\!]$). But it should be clear, and we shall show it later on, that most of what we do in this section can be applied to richer languages, containing also atomic formulae of the form $x \succ y$ or of any other form (with any number of variables).

The following deduction expressions assert that proximates are substitutable for each other in the two places of every atomic dominance formula ([LS] stands for left substitution and [RS] for right substitution).

[LS] $\quad x \sim y \vdash x \succsim z \rightarrow y \succsim z$

[RS] $\quad x \sim y \vdash z \succsim x \rightarrow z \succsim y$.

In a model $\underline{A}$ of the theory P of proximity these deductions translate to uniform continuity conditions on $[\![\succsim]\!]$ w.r.t. $[\![\sim]\!]$. Such an $\underline{A}$ is a model of [LS] and [RS] if for every ε there are δ_1 and δ_2 such that for every a,b,c in the domain A of $\underline{A}$ we have (using symmetry of proximity to obtain the absolute-value signs)

$\underline{[LS]}$ $\quad [\![a \sim b]\!] < \delta_1 \Rightarrow |[\![a \succsim c]\!] - [\![b \succsim c]\!]| < \varepsilon$

$\underline{[RS]}$ $\quad [\![a \sim b]\!] < \delta_2 \Rightarrow |[\![c \succsim a]\!] - [\![c \succsim b]\!]| < \varepsilon$.

What we have in $\underline{A}$ is an approximate substitutability, exactly as one should expect when dealing with proximity (an approximation to equality). This approximate substitutability, translating to uniform continuity, has some interesting interpretations and extensions which we now want to study through a series of theorems.

In the first theorem, if x and y are variables of L, φ a formula of L with variables $x_1, \ldots, x_n$ and $1 \leq i \leq n$, we denote by $\varphi(x_i/x)$ and $\varphi(x_i/y)$ the formulae of L obtained from φ by substituting x (y, respectively) for every occurrence of x_i in φ. If $\bar{a}$ is a function from $\{x, y, x_1, \ldots, x_n\}$ into the domain A of a structure $\underline{A}$ for L, s.t. $\bar{a}(x) = a$ and $\bar{a}(y) = b$ for some a and b in A, we set

$$\varphi\bar{a}(i/a) = \varphi(x_i/x)\bar{a}$$

$$\varphi\bar{a}(i/b) = \varphi(x_i/y)\bar{a} .$$

Theorem 1: If the structure $\underline{A}$ for L is a model of the theory PD of proximity and dominance, then, for every formula φ of L (with variables $x_1,\ldots,x_n$), for any variables x and y of L and for each i s.t. $1 \leqslant i \leqslant n$, $\underline{A}$ is a model of the substitutability deduction

$$[SU] \qquad x \sim y \vdash \varphi(x_i/x) \rightarrow \varphi(x_i/y).$$

In other words, for each i s.t. $1 \leqslant i \leqslant n$ and for every ε there is a δ s.t. for every function $\bar{a}$ as above

$$[\underline{SU}] \qquad [\![a \sim b]\!] < \delta \Rightarrow |\varphi\bar{a}(i/a) - \varphi\bar{a}(i/b)| < \varepsilon .$$

Proof: By induction on the structure of φ. If φ is atomic of the form $x_1 \gtrsim x_2$, then [SU] is just [LS] (with z standing for x_2) plus [RS] (with z standing for x_1). Then $\underline{A}$ is a model of [SU] by virtue of its being a model of dominance transitivity (see [DT] and [RT] in Section 2), of strong reflexivity (see [SR] in Section 3) and of symmetry (see [SY] in Section 2). If φ is atomic of the form $x_1 \sim x_2$ then [SU] is

$$x \sim y \vdash x \sim z \rightarrow y \sim z$$

$$x \sim y \vdash z \sim x \rightarrow z \sim y$$

(where again we write z for x_2 in the first deduction and for x_1 in the second). And clearly $\underline{A}$ is then a model of [SU] since it is a model of symmetry and transitivity ([SY] and [TR] of Section 2). In fact, in this case it follows from [$\underline{TR}$] that in the continuity condition [$\underline{SU}$] we can take $\delta = \varepsilon$. We shall use this fact, as well as the arguments in the following paragraph, in the proofs of the next two theorems.

To complete the proof of the present theorem let us assume that it holds for the formulae ψ and θ of L. Then for a given ε we can find $\delta_1,\delta_2,\delta_3,\delta_4$ s.t. $\underline{A}$ satisfies [$\underline{SU}$] with

(1) δ_1, ψ and ε

(2) δ_2, θ and ε

(3) δ_3, ψ and $\varepsilon/2$

(4) δ_4, θ and $\varepsilon/2$.

And now, to see that $\underline{A}$ satisfies [SU] with the given ε and with formulae built up from ψ and θ take

(i) δ_1 for $\neg\psi$

(ii) $\min\{\delta_1,\delta_2\}$ for $\psi \wedge \theta$ and $\psi \vee \theta$

(iii) $\min\{\delta_3,\delta_4\}$ for $\psi \rightarrow \theta$. □

The theorem just proved entails a strong continuity condition for (the interpretation of) φ in $\underline{A}$. Since φ is a formula with n variables in L, its interpretation in $\underline{A}$ is a [0,1]-valued function on A^n. Alternatively, for each variable x_i of φ we can say that we have a family $\varphi(x_i)$ of [0,1]-valued functions on A. Each member of the family is determined by a function $\bar{a}$ from the variables of φ into A, the values of $\bar{a}$ at variables other than x_i being parameters for this particular member of $\varphi(x_i)$. The theorem then tells us that the family $\varphi(x_i)$ is equi-continuous (i.e., the choice of δ for a given ε is independent not only of the particular value of $\bar{a}(x_i)$ but also of the values of $\bar{a}$ at other variables of φ.) The next theorem will tell us that this equi-continuity condition for $\varphi(x_i)$ at every i s.t. $1 \leqslant i \leqslant n$ implies uniform continuity of φ over A^n with the usual extension of the metric $[\![\sim]\!]$. (The opposite implication is obvious).

We shall now use the following notations. If $x_1,\ldots,x_n$, $y_1,\ldots,y_n$ are variables of L and A is the domain of a structure $\underline{A}$ for L then

$$\bar{a} : \{x_1,\ldots,x_n\} \rightarrow A$$

$$\bar{b} : \{y_1,\ldots,y_n\} \rightarrow A$$

$$\bar{x} \sim \bar{y} = \{x_i \sim y_i \ : 1 \leqslant i \leqslant n\}$$

$$[\![\bar{a} \sim \bar{b}]\!] = \max\{[\![\bar{a}(x_i) \sim \bar{b}(y_i)]\!] \ : 1 \leqslant i \leqslant n\}.$$

In the meta-language $\rightarrow$ denotes a function (not an implication). If φ is a formula of L with n variables we denote by $\varphi\bar{x}$ and $\varphi\bar{y}$ the formulae of L obtained from φ by substituting x_i (y_i, respectively) for each occurrence in φ of its i-th variable (for every i s.t. $1 \leqslant i \leqslant n$). We shall write $\varphi\bar{a}$ for $\varphi\bar{x}\bar{a}$ and $\varphi\bar{b}$ for $\varphi\bar{y}\bar{b}$.

Theorem 2: If the structure $\underline{A}$ for L is a model of the theory PD of proximity and dominance then, for every formula φ of L (with n variables for some positive integer n) and for any two sets $\{x_1,\ldots,x_n\}$ and $\{y_1,\ldots,y_n\}$ of variables of L, $\underline{A}$ is a model of the deduction

$$[SU]^* \qquad \bar{x} \sim \bar{y} \vdash \varphi\bar{x} \to \varphi\bar{y} \ .$$

In other words, for every ε there is a δ s.t. for each $\bar{a}$ and $\bar{b}$ as above

$$[\underline{SU}]^* \qquad [\![\bar{a} \sim \bar{b}]\!] < \delta \Rightarrow |\varphi\bar{a} - \varphi\bar{b}| < \varepsilon \ .$$

Proof: Consider the case of atomic φ's of the form $x \gtrsim y$. In this case [SU]* reduces to the deduction

$$x_1 \sim y_1,\ x_2 \sim y_2 \vdash x_1 \gtrsim x_2 \to y_1 \gtrsim y_2,$$

of which $\underline{A}$ is a model if for every ε we can find a δ s.t. for all a,b,c,d in the domain A of $\underline{A}$

$$\max\{[\![a \sim b]\!],[\![c \sim d]\!]\} < \delta \Rightarrow |[\![a \gtrsim c]\!] - [\![b \gtrsim d]\!]| < \varepsilon \ .$$

To see that this is so indeed, set $\delta = \min\{\delta_1,\delta_2\}$, after using [$\underline{LS}$] and [$\underline{RS}$] to find δ_1 and δ_2 s.t.

$$[\![a \sim b]\!] < \delta_1 \Rightarrow |[\![a \gtrsim c]\!] - [\![b \gtrsim c]\!]| < \varepsilon/2$$

$$[\![c \sim d]\!] < \delta_2 \Rightarrow |[\![b \gtrsim c]\!] - [\![b \gtrsim d]\!]| < \varepsilon/2 \ .$$

The proof for the case of atomic formulae of the form $x \sim y$ is similar. In fact it is simpler, since here, for any given ε, we can set $\delta = \varepsilon/2$, by the remark closing the first part in the proof of the preceding theorem (though we cannot take $\delta = \varepsilon$ as in that remark).

It remains to extend the claim of the present theorem to compound formulae; but this obviously follows from the induction in the second part of the proof of the preceding theorem. □

Note that the reason we took in the proof above the minimum between two δ's, each making a certain absolute difference smaller than $\varepsilon/2$, is that we are dealing here with 2-place atomic formulae. For an n-place atomic formula, in a language richer than L, we may have to choose between up to n δ's, each making some absolute difference smaller than ε/n (and we shall refer to this remark later on).

The following theorem provides an algorithm for finding a δ such that the error in certain substitutability assertions remains in a given ε-range whenever the error in some proximity assertions is below that δ.

Theorem 3: Let the structure A for L be a model of the theory PD of proximity and dominance, and for a given ε let $\delta_1(\varepsilon)$ and $\delta_2(\varepsilon)$ be the δ's with which A satisfies [LS] and [RS] for that ε. Then, for any ε and any formula φ of L with m implication signs (for some non-negative integer m), A satisfies [SU]* with

$$\delta = \min\{\delta_i(\varepsilon/2^{m+1}) : i = 1,2\}.$$

Proof: Without loss of generality, for any ε and for $i = 1,2$ we can assume $\delta_i(\varepsilon) \leqslant \varepsilon$. Thus, if we want A to satisfy [SU]* for φ and a given ε, we have to choose, for some ε', between the $\delta_1(\varepsilon')$ and $\delta_2(\varepsilon')$ appropriate for atomic components of φ of the form $x \gtrsim y$ (i.e., appropriate for [LS] and [RS]), and we can ignore δ's corresponding to components of the form $x \sim y$ since these, as noted above, can be set to equal ε'.

Now, by the proof of the preceding theorem and the remark following it, to find the appropriate ε' we first have to divide the given ε (with which we want A to satisfy [SU]* for φ) by 2. Then, each implication in φ may force a reduction of ε by another factor of 2, according to clauses (3), (4) and (iii) at the end of the proof of Theorem 1 above. The remaining clauses there tell us that other connectives do not force a change in the size of ε. □

For a fixed ε (and a fixed dependence of δ_1 and δ_2 on the varying ε) the theorem above provides a universal estimate of the δ required for the model A of PD to satisfy [SU]* for every formula φ with n variables and m arrows. As such, this is the "best" estimate for the "worst" among these φ's. For most of these φ's, however, we do not necessarily need such a small δ; we can choose between δ_1 and δ_2 for some ε' greater than $\varepsilon/2^{m+1}$. For instance, given such a φ, instead of m in the power of 2 at the denominator we can take the number of arrows in the longest chain of "nested" implications in φ. (This power can further be reduced by counting only those implications where at least one of the variables of L on one side of the arrow is identical to one of those on the other side. And still other reductions are possible in many cases).

But we note the following. First, for "short" formulae (let us say, for φ's with up to four or five connectives) the difference between the best possible δ and the estimate obtained from the theorem should not be too big. Second, the theorem is not meant to produce the best estimate of the required δ in every case, but rather to make the point that for any φ and ε we can determine a δ in advance if we know how δ_1 and δ_2 of [LS] and [RS] depend on ε in our structure $\underline{A}$. Thus the error in substitutability assertions is controllable in the strong sense that theorists, wishing to set a certain limit, ε, to this error for all formulae with a given number of implication signs, know in advance the level of precision, δ, they must achieve for proximity assertions (and so they can aim at the finest precision their "instruments" allow for a specific structure $\underline{A}$). In the following theorem we illustrate this controllability, in the simple case where the precision levels in a proximity assertion and in a corresponding substitutability assertion for any place of an atomic formula of L can be identical to each other.

Theorem 4: If the structure $\underline{A}$ for L is a model of the theory P of proximity and of the axiomatic versions of [LS] and [RS],

$$\vdash x \sim y \rightarrow (x \gtrsim z \rightarrow y \gtrsim z)$$

$$\vdash x \sim y \rightarrow (z \gtrsim x \rightarrow z \gtrsim y) ,$$

then, for any ε and for any formula φ of L with m implication signs (for some non-negative integer m) $\underline{A}$ satisfies [SU]* with

$$\delta = \varepsilon/2^{m+1} .$$

Proof: This is just a special case of the preceding theorem, where for any ε and for each of the two places of any atomic formula of L (whether of the form $x \sim y$ or of the form $x \gtrsim y$), $\underline{A}$ satisfies [SU] with $\delta = \varepsilon$. □

Differently put, this theorem asserts that, under its special condition, every formula φ of L with n variables (for some positive integer n), when interpreted as a [0,1]-valued function on the n'th power of the domain A of $\underline{A}$, satisfies the Lipschitz Condition. That is to say, there is a number p (specifically, $p = 2^{m+1}$, where $\underline{m}$ is the number of arrows in φ) s.t. for every $\bar{a}$ and $\bar{b}$, using the notations of Theorem 2 above,

$$p[\![\bar{a} \sim \bar{b}]\!] \geqslant |\varphi\bar{a} - \varphi\bar{b}| \ .$$

It is interesting to consider one more theorem, dealing with another version of substitutability (weaker, as we shall see, than [SU]*).

Theorem 5: If the structure $\underline{A}$ for L is a model of the theory PD of proximity and dominance then for every formula φ of L, using the terminology and notations of Theorem 2, $\underline{A}$ is a model of the deduction

$$[SU]^{**} \qquad \bar{x} \sim \bar{y},\ \varphi\bar{x} \vdash \varphi\bar{y} \ .$$

In other words, for every ε there is a δ s.t. for all appropriate $\bar{a}$ and $\bar{b}$

$$[\underline{SU}]^{**} \qquad \max\{[\![\bar{a} \sim \bar{b}]\!], \varphi\bar{a}\} < \delta \Rightarrow \varphi\bar{b} < \varepsilon \ .$$

Proof: Since $\underline{A}$ is a model of PD it satisfies [$\underline{SU}$]* for every φ and ε (with suitable δ's). Thus if for a given ε and a given φ we choose a $\delta(\leqslant \varepsilon/2)$ s.t. [$\underline{SU}$]* holds in $\underline{A}$ for φ with $\varepsilon/2$, it is easy to see that [$\underline{SU}$]** holds in $\underline{A}$ for φ with ε. □

This proof does more than verify the theorem. It shows that in any structure $\underline{A}$, which is a model of [SY] of Section 2, being a model of [SU]** follows from being a model of [SU]*. The converse is not necessarily true, and so [SU]** is usually weaker than [SU]* in such $\underline{A}$'s. We note that in ordinary two-valued logic [SU]* and [SU]** are equivalent and [SU]** is perhaps the more familiar version.

The structure $\underline{A}$ of this theorem is in particular a model of the following familiar versions of transitivity of proximity and transitivity of dominance:

$$x \sim y,\ y \sim z \vdash x \sim z$$

$$x \gtrsim y,\ y \gtrsim z \vdash x \gtrsim z \ .$$

To verify that $\underline{A}$ is a model of the first of these two deductions we can use [$\underline{TR}$] of Section 2 (which is, as we noted before, a strong form of a special case of [$\underline{SU}$]). For the second, which is not related to substitutability, we use [$\underline{DT}$] of Section 2 in exactly the same way. Each of the two deductions is, of course, weaker in $\underline{A}$ than the axiom, [TR] or [DT], used to establish $\underline{A}$ as a model of it.

Throughout this paper we restricted the language L to contain atomic formulae of proximity and dominance alone. As we noted at the beginning of Section 2 above (following Coombs [1964] and others), this should suffice for many, if not most, types of social and behavioral research data. However, we would like now, in the last part of this section, to consider briefly a more general language, L*, and extend to it the five theorems above.

We shall describe the language, and its structures, informally, then present without proofs the generalized theorems (followed by very brief remarks). Proofs for the generalizations of Theorems 1, 2, and 4 can be found in our [1982b]. The remaining two extensions are quite obvious and easy to prove.

In addition to 2-place formulae of the form $x \sim y$ and $x \gtrsim y$ the language L* contains a few (or many, perhaps even infinitely many) n-place atomic formulae for various positive integer n's. In addition to the connectives of L the language L* contains symbols for a universal and an existential quantifiers. Compound formulae are built up from connectives and quantifiers of L* in the usual way.

In a structure $\underline{A}$ for L*, with domain A, every n-place atomic formula of L* is interpreted as a [0,1]-valued function on A^n. Truth-values of atomic formulae of L* are determined by these interpretations in $\underline{A}$, and the truth-functionality of the connectives is as in clauses (3) to (6) of Section 2 above. For universally and existentially quantified formulae truth-values are determined by appropriate sups and infs, respectively. Axioms, deductions, axiomatic theories, deductive theories and the notions of structures being models of these, are defined in the case of L* exactly as in the case of L.

We turn now to the generalized theorems and we stipulate that the $\underline{A}$ appearing in each of them is a structure for L* (with domain A) and a model of the theory P of proximity (which, clearly, remains unchanged in L*). We use freely in L* the notations of all the preceding theorems.

<u>Generalization of Theorem 1</u>: If $\underline{A}$ is a model of [SU] for every atomic formula of L* then it is a model of [SU] for every formula of L*. (Of course, we speak here, both at the premise and at the conclusion, about being a model of [SU] at every variable x_i of the formula under consideration, as in Theorem 1 above. And we note that when seeking appropriate δ's for various ε's in the proof of this extended

theorem one finds that the presence of a quantifier may force a reduction in the size of ε, but not by more than a factor of 2).

<u>Generalization of Theorem 2</u>: For every formula φ of L*, <u>A</u> is a model of [SU] if and only if it is a model of [SU]*. (Again, when we speak about <u>A</u> being a model of [SU] for φ we mean for every variable x_i of φ. And in the proof here, as noted before, we may have to choose, for a given ε, between n δ's, each making a certain absolute difference smaller than ε/n).

<u>Generalization of Theorem 3</u>: Let φ be a formula of L* with k atomic components (for some positive integer k) and a total number of m implications and quantifiers (for some non-negative integer m). Let the j'th atomic component of φ be an n_j-place formula (for some positive integer n_j). If for every ε, every j s.t. $1 \leqslant j \leqslant k$ and every i s.t. $1 \leqslant i \leqslant n_j$, <u>A</u> satisfies [<u>SU</u>] at the i'th place of the j'th atomic component of φ with $\delta_{ij}(\varepsilon)$, then <u>A</u> satisfies [<u>SU</u>]* for φ and a given ε with

$$\delta = \min\{\delta_{ij}(\varepsilon/n_j 2^m) : 1 \leqslant j \leqslant k,\ 1 \leqslant i \leqslant n_j\}$$

(Once again we can obtain better estimates for δ but this does not alter the main point of the theorem).

<u>Generalization of Theorem 4</u>: If <u>A</u> satisfies the Lipschitz Condition (LC) for every atomic formula of L* then it satisfies LC for every formula of L*. (In particular, if the formula φ of L* has components, implications and quantifiers as in the generalization above of Theorem 3, then in the LC for φ we can have

$$p = \max\{p_{ij} n_j 2^m : 1 \leqslant j \leqslant k,\ 1 \leqslant i \leqslant n_j\}.$$

Here p_{ij} is a number with which <u>A</u> satisfies LC at the i'th place of the j'th atomic component of φ. That is to say, p_{ij} is s.t. for the i'th place of the j'th atomic component φ_j of φ we have

$$p_{ij}[\![a \sim b]\!] \geqslant |\varphi_j \bar{a}(i/a) - \varphi_j \bar{a}(i/b)|$$

for every $\bar{a}$ as in Theorem 1 above, but corresponding to φ_j).

Generalization of Theorem 5: For any formula φ of L*, if $\underline{A}$ is a model of [SU]* for φ then it is a model of [SU]** for φ. (The converse is not necessarily true, so [SU]** is weaker than [SU]* in $\underline{A}$).

5. DISCUSSION

In our controlled-error theories we assume that the error in an assertion φ can be controlled, i.e., kept below some ε of our own choice, by "making" certain assumptions true to within an appropriate δ. This assumption is stipulated in our interpretation, through the "logic of approximation" (LA; see Section 3 above), of deduction expressions of the form $\Gamma \vdash \varphi$. However, as is clear and known, in a multi-valued language like ours, with the same semantic rules for truth-values of compound formulae, other, more conventional, logics offer other interpretations of deduction expressions. We consider two such logics, which agree with LA on the interpretations of axioms but not on the interpretations of deductions from non-empty sets.

First, we mention what in [1980] and [1981c] we called "the logic of inexactness" (LI), which is essentially due to Scott [1974, 1976]. Under this logic, in a language L like ours or richer, a structure $\underline{A}$ for L with domain A is said to be a model of the deduction $\Gamma \vdash \varphi$, where φ is a formula of L and Γ a finite set of formulae of L, if for every function $\bar{a}$ from the set of variables of $\Gamma \cup \{\varphi\}$ into A

$$(\wedge\Gamma)\bar{a} \geqslant \varphi\bar{a} \ .$$

Second we recall what according to, e.g., Smiley's Comment on Scott [1976] is the "original" Lukasiewicz Logic (LL). Under this logic, with $\underline{A}$, Γ and φ as above, $\underline{A}$ is a model of the deduction $\Gamma \vdash \varphi$ if for every $\bar{a}$ as above

$$(\wedge\Gamma)\bar{a} = 0 \Rightarrow \varphi\bar{a} = 0 \ .$$

LI may be considered a special case of our LA, where we stipulate that, in the interpretation of deductions as given in Section 3 above, we can always set $\delta = \varepsilon$. In turn, LA is a special case of LL, where $\underline{A}$ is a model of $\Gamma \vdash \varphi$ if for every sequence $\{\bar{a}_n : 1 \leqslant n \lneq \infty\}$ of functions from the set of variables of $\Gamma \cup \{\varphi\}$ into the domain A of $\underline{A}$

$$\lim_{n\to\infty}(\wedge\Gamma)\bar{a}_n = 0 \Rightarrow \lim_{n\to\infty} \varphi\bar{a}_n = 0$$

(That this interpretation is equivalent to the ε - and - δ interpretation in Section 3 of this paper is proved in our [1982c]).

In this sense, then, our logic of approximation is between the orthodox Lukasiewicz Logic and the logic of inexactness. And not only in this sense, but also in the following one (derived from the first), which lies at the heart of this paper.

A deductive theory in LL is, in fact, not very different from what we called an error-free theory (or maybe we should call it an "ignored-error theory"), since we only consider the ideal case where the error on both sides of $\vdash$ is zero. A deductive theory in LI on the other hand, is exactly what we called an error-bound theory. It assumes that there are errors in both Γ and φ (and the error in Γ should always exceed that in φ if $\underline{A}$ is to be a model of $\Gamma \vdash \varphi$), without assuming that the theorist can control these errors. As a matter of fact, it is easy to see that being a model of the deduction $\Gamma \vdash \varphi$ in the sense of LI is equivalent to being a model of the axiom $\vdash \wedge \Gamma \rightarrow \varphi$ in the sense of our error-bound theories of Section 2 above.

So in this sense also, as we can now clearly see, in between LI and LL stands our LA, where errors exist, and are not ignored, but at the same time do not get out of control. This plausible property of LA has a few, not less plausible, consequences.

Consider, for instance, what is required from a formula ψ of L to be fully false in a structure $\underline{A}$ deduction-wise (i.e., to be s.t. $\psi \vdash \theta$ in $\underline{A}$ for every formula θ of L). In LI ψ is required to be everywhere semantically fully false (i.e., for every approximate $\bar{a}$ we must have $\psi\bar{a} = 1$). In LL it is enough that ψ is nowhere fully true (i.e., $\psi\bar{a} \neq 0$ for each appropriate $\bar{a}$). In LA we need the intermediate condition that ψ never approaches full truth (along any sequence $\{\bar{a}_n : 1 \leqslant n < \infty\}$ as above).

Similarly, for any two formulae, ψ and θ, of L to be equivalent (up to deduction) in a given structure $\underline{A}$ for L, in LI ψ and θ must be s.t. $\psi\bar{a} = \theta\bar{a}$ for every appropriate $\bar{a}$. In LL ψ and θ must have zeros in the same $\bar{a}$'s. In LA they should approach zero along the same sequences.

We want to show now that w.r.t. to some well-known laws and theorems of mathematical logic LA is closer to LL than to LI. Here, a law is an assertion that every structure for L is a model of a certain set of deductions, while a theorem, $D_1 \Rightarrow D_2$, is an assertion that every structure for L which is a model of a deduction expression of the form D_1 is also a model of the corresponding deduction expression of the form D_2. To say that a law or a theorem holds in a certain logic is to say that it holds when the notion of being a model (of some deductions) is interpreted in the sense of this particular logic.

The familiar Laws of Excluded Middle and Non-Contradiction, in the axiomatic forms

$$\vdash \psi \vee \neg \psi$$

$$\vdash \neg(\psi \wedge \neg \psi)$$

for every formula ψ of L, which, being sets of axioms are interpreted in exactly the same way in each of the three logics discussed above, do not hold in any of them. They do hold in LL and LA, but not necessarily in LI, in the deductive forms

$$\neg(\psi \vee \neg \psi) \vdash \theta$$

$$\psi \wedge \neg \psi \vdash \theta$$

for every two formulae ψ and θ of L (and compare this with our remarks above about the notion of being fully false in each of the three logics).

Similarly, for any two formulae ψ and θ of L, the Law of Detachment (Modus Ponens) holds in LL and LA, but not in LI, in its usual form

$$\psi, \psi \rightarrow \theta \vdash \theta \quad .$$

For every set, $\Gamma \cup \{\psi, \theta\}$, of formulae of L, one direction of the Deduction Theorem, namely,

$$\Gamma, \psi \vdash \theta \Rightarrow \Gamma \vdash \psi \rightarrow \theta \quad ,$$

holds in LI but neither in LL nor in LA, while the opposite direction,

$$\Gamma \vdash \psi \rightarrow \theta \Rightarrow \Gamma, \psi \vdash \theta \quad ,$$

holds in LA and LL but not in LI.

So much for laws of Logic. Getting closer to the main business of this paper, we note that the extension of substitutability from atomic to compound formulae of L (Theorems 1 and 2 and their Generalizations in the preceding section) holds in LA and LL but not necessarily in LI (in particular it is liable to fail in LI for formulae containing implication signs). However, the subtle way in which this extension is given above (including the algorithm for finding the appropriate δ's in Theorems 3 and 4 and their Generalizations) is peculiar to our LA. It tells us that errors exist and it shows us how to control them, i.e., how to keep them within a certain pre-determined range. (We note that Theorem 5 and its Generalization also hold in LA and LL but not in LI).

When we speak about substitutability in this paper it is always w.r.t. proximity, $\sim$, in L.

Under the logic of approximation it is interpreted in a model $\underline{A}$ of [SU] as uniform continuity w.r.t. $[\![\sim]\!]$. This assumes that $[\![\sim]\!]$ is a metric, and indeed it is preceded by an interpretation of $\sim$ as a (pseudo-) metric in a model $\underline{A}$ of the theory P of proximity, an interpretation which is common to the three logics above in view of the fact that [RE], [SY] and [TR] of Section 2 are axioms. In particular, transitivity, [TR], is interpreted as a triangle-inequality, and this is also what happens to dominance transitivity, [DT] (again in all three logics, as it is an axiom).

In the literature on proximity and dominance (and on the closely related indifference and preference) these notions are sometimes treated as two-valued relations (see, e.g., Scott and Suppes [1958]) and sometimes as multi-valued functions (see, e.g., Saati [1980] or Tversky [1974]). As relations they are sometimes transitive (as in, e.g., Luce and Raiffa [1957]) and sometimes intransitive (as in Roberts [1969] and in Tversky [1969]). These apparent "contradictions" are settled in our work, as explained in the preceding paragraph, by making proximity and dominance two-valued and transitive in the formal theories, but multi-valued and only approximately transitive in actual models of these theories.

In fact we provide a way of realizing the suggestion in Luce and Raiffa [1957] (p. 25) that axiomatic theories, ignoring errors, be viewed as theories of "normative" or "idealized" behavior, which is only an approximation to real-life behavior where errors and inconsistencies cannot be ignored. (In our logic, as we explained from the beginning, it is measurement error that leads from idealization in theories to approximation in models). Our approach also fits well with the fact that proximity (or similarity) is often studied through metric models (see, e.g., Gregson [1975]) where distances represent dissimilarities ("errors" in our terminology). This metric treatment is at the basis of the various multidimensional scaling techniques, not only the ones where similarity is quantitative (as in Torgerson [1958]) but also the so-called "non-metric" ones where similarity is merely ordinal (as in Rabinowitz [1975] and in Kruskal and Wish [1978]). And sometimes even dichotomous relations (e.g., indifference in Roberts [1969]) lead to (unidimensional) metric representations.

We note in conclusion that a similar "ambiguity" can be found in the literature on the modal interpretations of quantum logic. Accessibility (or similarity) of states, the central notion of these interpretations, is sometimes (e.g., in Goldblatt [1974] and in Dalla-Chiara [1977]) treated as a two-valued relation (it can be an equivalence relation if our modal system is the famous S5) and sometimes (as in, e.g., Bigelow [manu.]) as a multi-valued function, determined by a certain metric. In our [1982c] we proposed to settle this ambiguity in the way we did it here for proximity, and we used our interpretation of substitutability as uniform continuity, to axiomatize and analyze quantum propositions in the logic of approximation.

School of Education
Haifa University
Haifa, Israel.

REFERENCES

Bigelow, J.C.: Quantum Probabilities in Logical Spaces, Manuscript, Victoria University of Wellington.

Coombs, C.H.: 1964, _A Theory of Data_, John Wiley, New York.

Dalla Chiara, M.L.: 1977, Quantum Logic and Physical Modalities, _Journal of Philosophical Logic_ 6, 391-404.

Giles, R.: 1974, A Non-Classical Logic for Physics, _Studia Logica_ 33, 399-417.

Goldblatt, R.I.: 1974, Semantic Analysis of Orthologic, _Journal of Philosophical Logic_ 3, 19-35.

Gregson, R.A.M.: 1975, _Psychometrics of Similarity_, Academic Press, New York.

Katz, M.: 1980, Inexact Geometry, Notre-Dame Journal of Formal Logic 21, 521-535.

Katz, M.: 1981a, Difference Measurement Spaces, Journal of Mathematical Psychology 23, 195-213.

Katz, M.: 1981b, Lukasiewicz Logic and the Foundations of Measurement, Studia Logica 40, 209-225.

Katz, M.: 1981c, Two Systems of Multi-Valued Logic for Science, Proceedings of the 11th International Symposium on Multiple-Valued Logic, IEEE Computer Society Press, New York, 175-182.

Katz, M.: 1982a, Measures of Proximity and Dominance, Proceedings of the Second World Conference on Mathematics at the Service of Man, Universidad Politecnica de Las Palmas, 370-377.

Katz, M.: 1982b, Real-Valued Models with Metric Equality and Uniformly Continuous Predicates, Journal of Symbolic Logic 47, 772-792.

Katz, M.: 1982c, The Logic of Approximation in Quantum Theory, Journal of Philosophical Logic 11, 215-228.

Krantz, D.H., Luce, R.D., Suppes, P. and Tversky, A.: 1971, Foundations of Measurement, Academic Press, New York.

Kruskal, J.B. and Wish, M.: 1978, Multidimensional Scaling, Sage Publications, Beverly Hills, CA.

Luce, R.D. and Raiffa, H.: 1957, Games and Decisions, John Wiley, New York.

Rabinowitz, G.B.: 1975, An Introduction to Non-Metric Multidimensional Scaling, American Journal of Political Science 19, 343-390.

Roberts, F.S.: 1969, Indifference Graphs, in F. Harary (ed.) Proof Techniques in Graph Theory, Academic Press, New York, 139-146.

Saati, T.L.: 1980, The Analytic Hierarchy Process, McGraw-Hill, New York.

Scott, D.S.: 1974, Completeness and Axiomatizability in Many-Valued Logic, in L. Henkin et al. (eds.), Proceedings of the Tarski Symposium (Proceedings of Symposia in Pure Mathematics 25), American Mathematical Society, Providence, R.I., 188-197.

Scott, D.S.: 1976, Does Many-Valued Logic Have Any Use? in S. Körner (ed.), Philosophy of Logic, Basil Blackwell, Oxford, 64-95.

Scott, D.S. and Suppes, P.: 1958, Foundational Aspects of Theories of Measurement, Journal of Symbolic Logic 23, 113-128.

Shepard, R.N.: 1974, Representation of Structure in Similarity Data: Prospects and Problems, Psychometrika 39, 373-421.

Tarski, A.: 1956, Logic, Semantics, Metamathematics (papers from 1923-1938), Claredon Press, Oxford.

Torgerson, W.S.: 1958, Theory and Methods of Scaling, John Wiley, New York.

Tversky, A.: 1969, Intransitivity of Preferences, Psychological Review 76, 31-48.

Tversky, A.: 1977, Features of Similarity, Psychological Review 84, 327-352.

David Miller

IMPARTIAL TRUTH

My title, I must confess from the start, is something of a misnomer; at best, a halfhearted pun. It is not my intention today seriously to address the question of the impartiality or objectivity of truth, if only because it seems to me that little more need be said on this score. That truth in general is something that is beyond human decision, or human opinion (though not therefore something that is beyond human conjecture or human discovery), will be presumed in this paper to be a truth that is quite beyond any human desire to have it otherwise. (See, for example, Popper (1963), Chapter 10, and Popper (1972), (1979), Chapters 2 and 9; for a contrast see Goguen (1979).) Rather I wish to take up again the difficult, important, and at times depressing question of whether there can be any such animal as partial truth; whether in addition to truth and falsity themselves there are what might be called degrees of truth, or degrees of approximation to the truth; and if there are (or even if there are not), whether science has any need for such things. That is my topic - the partialness, or lack of it, of truth, not its partiality.

Although in a number of recent addresses (Miller (1980), (1981a), (1981b)) I have done my best to promote the view that we cannot do without degrees of truth in some way or another, today I intend to take a contrary tack. I wish to put before you a result, a result that has incidentally been in the literature for several years now (Miller, (1975); see also the interesting if impressionistic paper of Darmstadter (1975)), according to which we cannot persuasively make much sense of the idea that one scientific hypothesis is a better approximation to the true state of affairs than is another - unless, indeed, it happens actually to be true. Or to put the matter more modestly: if a sequence of hypotheses can constitute a sequence of finer and finer approximations to the truth, then the way in which that approximation is recognised is very different from what has normally been thought. And this spells trouble for our

H. J. Skala, S. Termini, and E. Trillas (eds.), Aspects of Vagueness, 75–90.

understanding of science, trouble that will not, I think, be dispelled by the many elaborate systems of many-valued logics, logics of approximation, fuzzy logics, and the like that have been generated in recent years to take account of the undoubted fact that our scientific predictions are quite often not quite right. For my attack is in effect an attack on what most, if not all, of these systems take for granted: namely that science is the exercise in successive approximation that its recent apologists (I mention here only Kuhn (1970), p.206, since he is also amongst its detractors) have always supposed it to be. I myself do not see how it can be this, though I would assuredly like to be mistaken.

Great strides, if not great advances, have been made in recent years towards the understanding of the progress of science. And we can now at least distinguish (as indeed Hilpinen (1976) did quite early on) between various ways in which we might hope our scientific work to be progressing towards the truth. In particular we may distinguish vertical and horizontal dimensions of improvement, in a manner that I shall explain. Vertical improvement consists largely in our attempts, feeble as they are, to grasp the whole truth (as it can be formulated in any given language); the improvement is an improvement at least partly in content, but mainly in truthlikeness or in verisimilitude (the terms are used synonymously). Now no one, I trust, even Tichý and Oddie at their most evangelical (Tichý (1978), Oddie (1981)), takes too seriously the extraordinary claim that the problem of verisimilitude has actually been solved - that it has been shown in what way Newton's gravitational theory, for example, is more truthlike than Galileo's (if it is). But verisimilitude is not today's business. Here we shall be concerned more with the possibilities for horizontal improvement in a scientific hypothesis. By this I most emphatically do not mean improvements in its scope at (say) the empirical level; for an increasing wealth of predictions, like an increase in truthlikeness, is a consequence of vertical improvement - which boils down, somehow or other, to an increase in logical and empirical content. No, what I am concerned with here is the question of fit: the way in which some hypotheses, not obviously richer in content than their rivals, nevertheless appear to fit the facts better - if you like, to permeate more insistently into the crevices of the empirical material. In short, the extent to which one hypothesis can improve in accuracy on another, or be a better approximation to the truth than is that other, is the issue that I wish to discuss again, and in a somewhat different way.

Granted, then, that questions about truthlikeness be left to one side, we may devote ourselves to the question of degrees of truth; or, as Scott (1974), p.421, and, following him, Katz (1981a), 1981b), (1984), call them, degrees of error or of inexactness. Scott asks plaintively in his (1976) whether there is a way of making sense of the many values to be found in the systems of many-valued logic. And in each of the papers mentioned he proposes that perhaps such additional truth values may indeed be understood as degrees of error, at least in the one-dimensional case ((1974), pp.421f.; (1976), pp.66-70). Katz too has investigated this case (for example in his (1980)). In this paper I am going to voice some intuitive, rather than technical, doubts whether even this one-dimensional case can be dealt with satisfactorily (see also the pertinent comments of Smiley (1976)), and some very strong technical doubts concerning the many-dimensional case (looked forward to, for example, by the reviewer of Katz (1980) in Mathematical Reviews MR 82c:03017). I cannot properly understand degrees of error in the many-dimensional case. Perhaps others can.

I am taking it for granted that we are not going to think of degrees of error as somehow lying between truth and falsity, as though there was anywhere there for them to lie. That kind of approach to the problem, in which truth and falsity figure simply as two amongst a lot, perhaps a continuum, of other values, is a decidedly unappealing one; still worse is the kind of approach in which truth and falsity disappear altogether, and the two-valued scale is not simply supplemented, but actually replaced, by a many-valued scale. Formally such systems may have their interest, but I do not believe that they are of the slightest interest to science. What we are concerned with in science, after all, is the truth about the world; with attaining it if we can, and with somehow approximating it if we do not attain it. And our method for doing this is one that searches for falsity. Any picture of science, therefore, in which falsity disintegrates as a real option, or worse, in which both truth and falsity disappear, is simply not true to the way we do science; in particular, it leaves in the deepest shadow the way in which the results of the empirical tests we perform are supposed to impinge on the hypotheses that we propose. To conclude this little diatribe, therefore, let me make clear that for me there are only two truth values: truth and falsity. Nevertheless, it may be possible to subdivide the lower of these (falsity) more or less finely - to recognise, that is, degrees of falsity, or of error, or of approximation to the truth. Whether we can actually achieve this is however a different matter.

Errors abound in the sciences, and are cheerfully admitted; and the reason seems to be that errors are not seen to be the same kind of things as mistakes. Whoever heard of a logic of mistakes? Indeed, it is just because of this that, I suspect, attempts have often been made to insinuate degrees of error between the more stately truth values of truth and falsity. Far and away the most important species of errors are, of course, numerical errors; and it is to these that I shall devote most of my attention. They provide, after all, in the one-dimensional case, as simple an example as you could want for the idea of a degree of error, or degree of inexactness. You measure some parameter f and obtain the value 2 ; I measure it and obtain the value 3 . In fact the true value is just 1 . It seems as clear as anything not only that my measurement is more in error than yours, but that it is twice as much in error. We have not only an ordering of errors, but a way of grading or measuring them. Equally we may say that your result approximates better to the truth - in the sense of approximating better to the true value of f - than does mine; indeed, that it is twice as good an approximation to the truth. Of course scale changes would lead to different evaluations of how much more in error my measurement is than yours; but this is just the sort of thing we expect from scale changes, and cannot be thought to pose a serious problem. Indeed, nothing the least contentious, or even interesting, appears to have been said so far. There is a clear and unambiguous sense, it seems, in which some evaluations of numerical parameters are more in error, and measurably more in error, than are some others.

Here already, however, I begin to see problems and difficulties. Suppose that, unbeknownst to us, we are not in fact measuring the parameter f but some related (though independent) parameter g. It is hard to credit that this does not sometimes happen, perhaps because there is some further factor that we have not accounted for. A simple example might be a Galilean measurement of the acceleration due to gravity that in fact measures (correctly or incorrectly) only the acceleration at a particular altitude. In such a situation it certainly looks as though there is, in addition to any strictly numerical error made, also a straightforward mistake - the mistake of supposing that it is f that we have measured rather than g. To put the matter in sentential terms, the measurement report 'f = 3' may be an incorrect report not only because 3 is the wrong value but because f is the wrong parameter. It may perhaps be felt that there is no difficulty here: given that f actually equals 1 the report that f = 3 is much more in error than the report that f = 2. That is fair enough. But the point I am trying to make is that it may not be clear which truth, or true sentence, it is that our measurement report is attempting to approximate. As an answer to the question 'What is the true value of f?' doubtless 'f = 2' is better than 'f = 3'. But as an answer to the question 'What is the value of g?' (which, in the context, is really the question that is being answered), it may not be so. More generally we may say that the extent to which a sentence is in error must depend on what question it has been proposed to answer, and, more especially, on what is supposed to be the true sentence at which it is aimed (some writers have referred in this context to the target). Now in the case of measurements of single parameters the matter is usually, though, as I have tried to argue, not universally, straightforward. But in most other cases, especially most of those of concern to theoretical science, straightforward it usually is not.

Examples abound of scientific hypotheses and theories that are admitted to be in error, and in some clear sense merely numerically in error, for which it would not be easy to say what the true target hypothesis was. Kepler's three planetary laws provide a striking instance. They are acknowledged to be incorrect. Yet what is the truth for which they are striving? It is surely not just a catalogue of all the planetary positions, a completion as it were of Brahe's data. For such a catalogue leaves out of account, apparently, many numerical interrelationships that Kepler's laws (rightly or wrongly) succeed in bringing to light. Thus quite apart from the complexity of the errors that Kepler's laws make, it seems to me to be very far from straightforward what the extent of those errors might be thought to be. In other words, it is very unclear that there is such a simple unadorned quantity as the degree of error, or degree of approximation to the truth, of Kepler's theory. Allowing ourselves some massively generous helping of idealization we may perhaps say that for any particular truth, that is, true sentence, Kepler's theory approximates that truth in some degree. To show that such things can be sensibly said is part of the purpose of my paper 'A Geometry of Logic' (1982), which is printed in the proceedings of this conference. But it does not follow from this that there is some degree to which Kepler's theory approximates the truth. Of course there is a degree to which it approximates the whole truth; that is what we call its truthlikeness or verisimilitude. But here, as I have already stressed, we are not concerned with a theory's shortcomings in content; only with its shortcomings in accuracy and in fit.

To summarize this, my first main point. Although the truth or falsity of a hypothesis is something that is quite independent of what questions that hypothesis was proposed as an answer to (provided we eliminate context-dependent elements, indexicals and the like), this cannot be claimed of the degree of truth, or degree of error of the hypothesis. Many writers have, I am afraid, incautiously generalized from the case of evaluations of single parameters, where the matter might be presumed to be uncontentious enough, to the general case. Only if one does so generalize, of course, is there any point in having a real logic of error; if we were only concerned with numerical evaluations we could presumably restrict ourselves to equational logic, or something like it, and the apparatus of sentential and predicate calculus could be dispensed with. But once we admit that the degree of error of a hypothesis is a relative matter - that is, is relative to what is regarded as the true sentence being approximated - we shall too allow that the degree of error of a sentential compound of two hypotheses may well be no function at all of the separate degrees of error; or if it is some function, that function is likely to be an intricate one. I mention this point not because I want to follow it up at all here, but only to emphasize that the very simple criticism so far made may have dramatic consequences for all those many attempts to formulate logics of error based in some manner on the original many-valued logics of sixty years ago.

Happily, I myself do not believe that any such logics are necessary, and am convinced that all questions concerning the approximation of one hypothesis by another can be sensibly accommodated within the classical tradition. My other paper at this conference, 'A Geometry of Logic', sketches one way in which this might be done. Now, however, I wish to strike more deeply at the presumption that every false hypothesis is true in some degree by questioning not just the absoluteness of this supposed degree of truth but its existence altogether, even in the numerical case. I am going to put before you a new and improved version of the result, first sketched in my (1975), that except in the one-dimensional case the degree of error, or degree of exactness, or degree of accuracy, of a scientific hypothesis is not invariant under trivial mathematical reformulations. More strongly, that even an ordering by relative accuracy is not so invariant. If we assess, as we are surely inclined to do, the accuracy of a prediction of the value of a parameter by the extent of its separation from the true value, then we shall find that in one formulation one hypothesis is uniformly more accurate and therefore presumably less in error, than another; whilst in another formulation, as physically meaningful as the first, the reverse obtains. This result is exceedingly general, and will be shown to apply in almost all cases of hypotheses that seek to predict the values of more than one parameter. If degrees of error exist for such hypotheses, therefore, they do not simply vary directly with the accuracy that the hypotheses achieve.

Elementary numerical illustrations are easier to cope with than any algebraic manipulations, so I shall start with an elegant example devised by Sir Karl Popper and presented in the new appendix of his (1979). Let A and B be two rival hypotheses engaged in predicting the values of the parameters f and g - and, we may suppose, nothing else. Suppose that their predictions, along with the values given by the true theory T, are as stated in the accompanying table:

	f	g
A	0.150	1.225
B	0.100	1.000
T	0	1.000

It then appears that with regard to both f and g the hypothesis B is superior in accuracy to the hypothesis A; and therefore, one assumes, is overall the less erroneous hypothesis. But now define in terms of f and g two new parameters h and k as follows:

$$h = g - 2f$$
$$k = 2g - 3f.$$

In terms of these new parameters the performance of A and B may be simply calculated, and stated thus:

	h	k
A	0.925	2.000
B	0.800	1.700
T	1.000	2.000

With regard to these parameters it is readily apparent that A is the more accurate, that is, less erroneous, hypothesis. So the relative degrees to which the hypotheses A and B are in error depends, it seems, not just on what they say but on how they say it. And there is not, on the face of it, much to choose here between the ways that they do say it. It is not as though h and k are a thoroughly contrived and unphysical pair of parameters, compared with which f and g are refreshingly direct. Or, rather, if this is the case I do not know why it should be: for it turns out that f and g may be redefined from h and k in just the way that h and k were defined from f and g:

$$f = k - 2h$$
$$g = 2k - 3h.$$

Nothing could be more deliciously symmetrical. In particular cases, of course, there may be substantial physical reasons for preferring one of these pairs of parameters to the other. But we all know that there are scores of physical parameters - lengths, distances, velocities, masses, and the like - linear combinations of which are just as meaningful parameters in their own right. Thus we can anticipate that there will be many instances where the degree of error of some simple empirical hypothesis is not straightforwardly to be determined from the approximate correctness of the predictions it makes on some selected set of parameters.

Reversals of ordering by accuracy can of course be obtained even in the one-dimensional case if we are prepared to allow discontinuous transformations (see Good (1975), p.205 and Tichý (1978), pp.194f., who somewhat remarkably makes no mention at all of my (1975)). But this can scarcely be regarded as interesting, as some topological restraints must be insisted on if our reformulated hypotheses are to be thought of as reformulations at all. What is interesting, surely, is that there is so much topological variance in the midst of such a concordat on continuity. Notice, incidentally, that the actual numerical values recorded above, are of little importance: what has been shown is that although B's predictions of the values of f and g lie between those of A and of T, in the case of the parameters h and k it is A's predictions that lie between those of B and T. The result is unaffected by any change of scale for some or all of the parameters. The illustration therefore points to something much more general, to a result that is fully ordertheoretic and depends on no specific numerical evaluations at all.

I now propose to show, briefly and I hope without excessive fuss, how such a generalization may be achieved. Originally such a proof was given in my (1975), but the proof there was lopsided in character, and inelegant around the edges. The present demonstration is much to be preferred, for its symmetry as well as for its conciseness. We shall suppose, as before, that f and g are numerical parameters, perhaps depending on independent variables s, t, ... (which we shall systematically suppress), and that A and B are hypotheses predicting the values of these parameters. That is to say there are numerals f_A, f_B, g_A, g_B such that $A \vdash f = f_A \;\&\; g = g_A$ and $B \vdash f = f_B \;\&\; g = g_B$.

As above we shall use T for the true theory, and f_T and g_T for the true values of f and g, so that $T \vdash f = f_T \; \& \; g = g_T$, and we shall construct two simple linear combinations of f and g in terms of which the hypotheses A and B behave in topsy-turvy ways; and for this purpose we will need further numerals p, q, μ. (It will be possible for p, q, μ, too to depend on the independent variable t.) Let us then write down our assumption, which is that with respect to the parameters f and g the hypothesis B is uniformly more accurate than is the hypothesis A:

$$\begin{array}{ll} \text{either} & f_A < f_B < f_T \text{ or } f_T < f_B < f_A \\ \text{and either} & g_A < g_B < g_T \text{ or } g_T < g_B < g_A \end{array} \qquad (1)$$

The case where some of these inequalities might become equalities (as in Popper's illustration) can be accommodated, but it is necessary that enough inequalities remain to ensure first that A and B are distinct hypotheses and second that B and T are distinct (that is to say, B is not true). As we shall see, we shall also need to require that f and g are genuinely distinct parameters, so that we do not collapse back to the one-dimensional case.

Now the condition that a linear combination f + pg should succeed in reversing A and B in accuracy; that is,

$$\begin{array}{ll} \text{either} & f_B + pg_B < f_A + pg_A < f_T + pg_T \\ \text{or} & f_T + pg_T < f_A + pg_A < f_B + pg_B, \end{array} \qquad (2)$$

may be written in the form

$$\text{if } f_B + pg_B \lessgtr f_A + pg_A \text{ then } f_A + pg_A \lessgtr f_T + pg_T. \qquad (3)$$

So whether or not $g_A < g_B$ or $g_B < g_A$ it follows that

$$\text{if } \frac{f_B - f_A}{g_A - g_B} \lessgtr p \text{ then } \frac{f_A - f_T}{g_T - g_A} \gtrless p. \qquad (4)$$

We may therefore choose for p some weighted average of these two fractions, say

$$p = \mu \cdot \frac{f_B - f_A}{g_A - g_B} + (1 - \mu) \frac{f_A - f_T}{g_T - g_A}, \qquad (5)$$

where $0 < \mu < 1$. Provided the two fractions being weighted are not actually equal,

$$\frac{f_B - f_A}{g_A - g_B} \neq \frac{f_A - f_T}{g_T - g_A}, \qquad (6)$$

in which case we would have equality almost everywhere, we can proceed to define a new parameter h by

$$h = f + pg, \qquad (7)$$

and with respect to h the hypothesis A will be more accurate than the hypothesis B. In a similar way we can write

$$q = \mu \cdot \frac{g_B - g_A}{f_A - f_B} + (1 - \mu) \cdot \frac{g_A - g_T}{f_T - f_A} \qquad (8)$$

and define a new parameter k by

$$k = g + qf .$$

In terms of h, k we may then redefine f, g by noting that

$$\begin{aligned} f &= h - pg \\ &= h - pk + pqf , \end{aligned} \qquad (9)$$

so that,

$$f = \frac{pk - h}{pq - 1}, \quad g = \frac{qh - k}{pq - 1} . \qquad (10)$$

A highly symmetrical set of definitions is obtained with

$$h = \frac{pg + f}{\sqrt{pq - 1}} \qquad k = \frac{qf + g}{\sqrt{pq - 1}} \qquad (11)$$

$$f = \frac{pk - h}{\sqrt{pq - 1}} \qquad g = \frac{qh - k}{\sqrt{pq - 1}} \tag{11}$$

where pq is the product of the terms on the right-hand sides of equations (5) and (8). It can be shown that $pq \geq 1$; and that it equals 1, thus inciting a discontinuity somewhere, if and only if the condition (6) fails; that is to say,

$$\frac{f_B - f_A}{f_A - f_T} = \frac{g_B - g_A}{g_A - g_T} \tag{12}$$

Equation (12) will happen, for example, when f is some linear combination of g, and in other cases too where their independence as parameters will seriously be in question.

Given that (12) does hold, our construction will be unsuccessful. Nevertheless it is usually easy enough to provide an alternative that will do the trick. We may, for example, if f and g are non-negative, replace f and g by f^2 and g^2. Our assumption that B is more accurate than A will not be affected by this substitution. Now the analogous condition to (12), namely

$$\frac{f_B^2 - f_A^2}{f_A^2 - f_T^2} = \frac{g_B^2 - g_A^2}{g_A^2 - g_T^2}, \tag{13}$$

will hold together with (12) only if

$$\frac{f_T}{g_T} = \frac{f_A}{g_A} = \frac{f_B}{g_B}; \tag{14}$$

that is, only if f is a numerical multiple of g. In such a circumstance it cannot be said that within the context of the hypotheses A and B the parameters f and g are seriously distinct.

To illustrate the construction let us take a simple example where each of f and g is a function of an independent variable t. The predictions issued by A, B and T are as follows:

	$f(t)$	$g(t)$
A	t	$5mt$
B	$t+a$	$2mt$
T	$t+2a$	mt

where a and m are constants, or, anyway, are not themselves functions of t. (It is assumed that the constant a is of the same dimension as t.) With respect to the parameters $f(t)$ and $g(t)$ the hypothesis B is manifestly more accurate than the hypothesis A. Yet trivial calculations show that $p = 5a/12mt$, that $q = 5mt/2a$, and thus that $pq - 1 = 1/24$. The new parameters take a form that is only deceptively complicated:

$$h(t) = \frac{5a\sqrt{24}}{12mt}\, g(t) + \sqrt{24}.f(t) \tag{15}$$

$$k(t) = \frac{5mt\sqrt{24}}{2a}\, f(t) + \sqrt{24}.g(t)$$

and the predictions made for them (after division throughout by $\sqrt{24}$) are as follows:

	$h(t)/\sqrt{24}$	$k(t)/\sqrt{24}$
A	$t + 25a/12$	$5mt^2/2a + 5mt$
B	$t + 22a/12$	$5mt^2/2a + 9mt/2$
T	$t + 29a/12$	$5mt^2/2a + 6mt$

As was to be expected, A is the more accurate hypothesis for each of h and k throughout the whole range of t.

Our construction, though mathematically elementary, is obviously of considerable generality. Yet as Horton (1978) has pointed out, it is not as radical as it appears at first sight. For there are precious few examples in the history of science where one competing hypothesis turns out to be uniformly more accurate than another, even on a single parameter; and in most real cases, therefore, other considerations (such as functional form) have to be brought to bear if we are to express any empirical preference between the competing hypotheses. And the case that I have considered can be seen as just a special instance of this much more general phenomenon. Though for the parameter f the hypothesis B may be uniformly more accurate than A over the range of t, when the weighting factor μ is allowed to vary as well a very different picture emerges. I cannot go into Horton's discussion in any detail here; I shall merely emphasize, as he does, that though the result may be less radical it is not therefore any less serious. For just as there is no such relation between hypotheses of one's being uniformly the more accurate, so there is no relation of one's being preponderantly the more accurate. From a formal point of view false hypotheses that predict values for the same set of parameters are on a par. (There is another difficulty here that I pass over; see my (1975), pp.166f.) If there is a reason for preferring one to another it must constitute a reason for preferring one family of parameters to another. In my (1975), pp.181-186, I expressed considerable doubts that there could be such reasons, anyway in the general case. What apologists for the idea that our scientific hypotheses do approximately explain the facts (such as Girill (1978)), or that they approach the truth through successive approximation or observational nesting (such as Newton-Smith (1981) pp.206f.), must surely do is to come forward with some such reasons. One cannot simply disregard what is, after all, a plain and straightforward result of something like high-school mathematics.

Nothing that I have said here proves, of course, that there cannot be degrees of error, or degrees of inexactness, or (looking at it from the other end) degrees of truth. All that can be said is that, if they exist, they depend on factors of which we have at present little cognisance. If only they did exist, one is inclined to cry. For from a methodological point of view their presence (and in even some sense calculability) is of the most considerable importance. There is no doubt, after all, that we often act in the light of false hypotheses; but what rational basis there can be for this remains a mystery if we cannot depend on some such rule as that hypotheses that are approximately true have consequences that are themselves approximately true (a slogan that, as we have seen, is demonstrably false when numerical accuracy is substituted for approximate truth.) This I have discussed in greater detail in my (1980), and especially in (1981a) and (1981b). Despite the insipid discussion of Haack (1980), especially pp.15-17, therefore, there seems to me to be the strongest possible methodological demand for our being able to make sense in some way of the degree to which some hypothesis, in answer to some problem, approaches what might be thought to be the true answer to that problem. I am no less sorry than I was ten years ago that I seem unable to do much at all to satisfy that demand.

Department of Philosophy,
University of Warwick

NOTE

This paper has also been published in Stochastica 6, 1982, pp. 169-186. I am much indebted to Mrs Jean Nightingale for her skilful typing of the paper.

REFERENCES

Darmstadter, H.: 1975, 'Better Theories', Philosophy of Science 42, 20-27.

Girill, T.R.: 1978, 'Approximative Explanation', in P.D. Asquith and I. Hacking (eds.), PSA 1978, volume 1, Philosophy of Science Association, East Lansing, pp. 186-196.

Goguen, J.A.: 1979, 'Fuzzy Sets and the Social Nature of Truth', in M.M. Gupta, R.K. Ragrade, & R.R. Yager (eds.), Advances in Fuzzy Set Theory and Applications, North-Holland Publishing Company, Amsterdam, pp. 49-67.

Good, I.J.: 1975, 'Comments on David Miller', Synthese 30, 205f.

Haack, S.W.: 1980, 'Is Truth Flat or Bumpy?', in D.H. Mellor (ed.), Prospects for Pragmatism, CUP, pp. 1-20.

Hilpinen, R.: 1976, 'Approximate Truth and Truthlikeness', in M. Przełęcki, K. Szaniawski, & R. Wójcicki, (eds.), Formal Methods in the Methodology of Empirical Sciences, Reidel, Dordrecht, & Ossolineum, Wrocław, pp. 19-42.

Horton, R.J.: 1978, 'Accuracy of Prediction: A Note on David Miller's Problem', British Journal for the Philosophy of Science 29, 179-183.

Katz, M.: 1980, 'Inexact Geometry', Notre Dame Journal of Formal Logic 21, 521-535.

Katz, M.: 1981a, 'Two Systems of Multi-valued logic for Science', Proceedings of the 11th International Symposium on Multiple-valued logic, 175-182.

Katz, M.: 1981b, 'Łukasiewicz Logic and the Foundations of Measurement', Studia Logica 40, to appear.

Katz, M.: 1984, 'Real-valued Models with Metric Equality and Uniformly Continuous Predicates', Journal of Symbolic Logic 47, 772-792.

Kuhn, T.S.: 1970, The Structure of Scientific Revolutions, University of Chicago Press, second edition.

Miller, D.W.: 1975, 'The Accuracy of Predictions', Synthese 30, 159-191.

Miller, D.W.: 1980, 'Verisimilitude in Action', presented at the Symposium on the Philosophy of Sir Karl Popper, London, 1980.

Miller, D.W.: 1981a, 'Vérité, Vérisimilitude, Véridicité', presented at the colloquium, Karl Popper et la science d'aujourdhui, Cérisy-la-Salle, 1981.

Miller, D.W.: 1981b, 'Truth, Truthlikeness, Approximate Truth', English version of 1981a, forthcoming in Fundamenta Scientiae.

Miller, D.W.: 1982, 'A Geometry of Logic', Proceedings of the Second World Conference on Mathematics at the Service of Man, Las Palmas, 1982. Reprinted in the present volume.

Oddie, G.: 1981, 'Verisimilitude Reviewed', British Journal for the Philosophy of Science 32, 237-265.

Popper, K.R.: 1963, Conjectures and Refutations, Routledge & Kegan Paul, London.

Popper, K.R.: 1972, Objective Knowledge, OUP.

Popper, K.R.: 1979, Objective Knowledge, OUP, second edition.

Scott, D.S.: 1974, 'Completeness and Axiomatizability in Many-Valued Logic', Proceedings of the Tarski Symposium (Proceedings of Symposia in Pure Mathematics XXV), American Mathematical Society, Providence, pp. 411-435.

Scott, D.S.: 1976, 'Does Many-valued Logic Have Any Use?', in S. Körner (ed.), Philosophy of Logic, Blackwell, Oxford, pp. 64-74.

Smiley, T.J.: 1976, Comment on Scott (1976), ibid., pp.74-88.

Newton-Smith, W.H.: 1981, The Rationality of Science, Routledge & Kegan Paul, London.

Tichý, P.: 1978, 'Verisimilitude Revisited', Synthese 38, 175-196.

David Miller

A GEOMETRY OF LOGIC

This paper[1] will sketch a theory of partial identity or partial congruence for the elements of a Boolean algebra, and will establish the equivalence of this theory with the theory of metric Boolean algebras and with that of unnormalized measures. (It is not suggested that this latter equivalence is not well known.) The merit of the theory, and of others like it, is that it allows us, with no departure from classical two-valued logic, to contemplate how scientific hypotheses may approximate to each other, as well as to the truth. The theory also suffers from some disadvantages, which will be noted.

The letters a, b, c will be used for the elements of an arbitrary Boolean algebra. The join and meet operations of the algebra will be represented by + and by concatenation respectively. The complementation operation will be represented by an accent. The symmetric difference of a, c, defined by

$$a \Delta c \;=\; ac' + a'c,$$

will play a crucial role in much that we do. Where necessary, the ordering relation of the algebra may be represented by $\leq$ or by $\geq$, and its zero and unit by 0 and 1.

Some of these symbols, in particular +, 0, $\leq$, and $\geq$, not to mention =, will be used additionally in reference to some partially ordered abelian semigroup with unit, normally the non-negative real numbers, which will act as the space of values for a function d defined on the Boolean algebra. There should not be any real possibility of confusion between the different uses of these symbols. At one or two places, which will be noted, it will be possible to generalize our results if we assume some cancellation principle for the semigroup.

1. PARTIAL IDENTITY

The relation of identity is usually characterized by the axioms

H. J. Skala, S. Termini, and E. Trillas (eds.), Aspects of Vagueness, 91–104.

$$a = a,$$
$$a = c \ \& \ \Phi a \rightarrow \Phi c,$$

where ϕa is a formula perhaps involving a and Φc is like it except perhaps for containing c free at one or more places where Φa contains a free. If we write $\bot$ for the falsum or absurdity, and $\Psi \leftarrow \Theta$ for $\Theta \rightarrow \Psi$ we may accordingly characterize also the relation $\neq$ of diversity by the axioms

$$a \neq a \leftrightarrow \bot,$$
$$a \neq c \lor \neg \Phi a \leftarrow \neg \Phi c.$$

In the case of algebraic languages, where = is the only relation symbol, the second principle, of the indiscernibility of identicals, mav be put in the form

$$a \neq c \lor \phi a \neq b \leftarrow \phi c \neq b,$$

where ϕa is here a term perhaps involving a and the conventions on substitution are as before. In developing a theory of the partial identity or partial congruence of two elements of an algebra we shall seek to generalize these axioms in terms of a (typically real-valued) function d that satisfies

(0) $\quad d(a, a) = 0$

and

(1) $\quad d(a, c) + d(\phi a, b) \geq d(\phi c, b).$

In the presence of the definition of identity for Boolean elements

$$a = c =_{Df} d(a, c) = 0$$

the axioms (0) and (1) are easily seen to generalize the diversity axioms. They are axioms for dissimilarity. We shall here explore their consequences.

Consider first a degenerate algebra with no operations. The only term that involves a will be a itself, and so we shall extract from (0) and (1) no more than

(2)	$d(a, c) + d(a, a) \geq d(c, a)$	1
(3)	$d(a, c) = d(c, a)$	0,2
(4)	$d(a, c) + d(a, b) \geq d(c, b).$	1

In other words, d is a pseudometric operation.

If our algebra does contain operations, they will be subject to a rather strong condition of uniform continuity:

(5)	$d(a, c) + d(\phi a, \phi a) \geq d(\phi c, \phi a)$	1
(6)	$d(a, c) \geq d(\phi a, \phi c)$	0,3,5

It turns out that (6) is easier to work with than (1) is. It is worth noting that from (6) and the triangular inequality (4) it is possible to derive (1).

We consider now the introduction of the lattice operations.

(7)	$d(a, c) \geq d(b + a, b + c)$	6
(8)	$d(a, c) \geq d(ba, bc)$	6
(9)	$d(a, ac) \geq d(c + a, c + ac)$	7
(10)	$d(a, ac) \geq d(a + c, c)$	9
(11)	$d(a + c, c) \geq d(a(a + c), ac)$	8
(12)	$d(a + c, c) \geq d(a, ac)$	11
(13)	$d(a, ac) = d(a + c, c)$	10,12

The lattice quadrangles thus are parallelograms.

We now explicitly assume that the lattice we are dealing with is a Boolean algebra, with the symmetric difference defined as above. The derivations continue:

(14)	$d(a, c) \geq d(a \triangle c, c \triangle c)$	6
(15)	$d(a, c) \geq d(a \triangle c, 0)$	14
(16)	$d(a \triangle c, 0) \geq d(c \triangle (a \triangle c), c \triangle 0)$	6
(17)	$d(a \triangle c, 0) \geq d(a, c)$	16
(18)	$d(a, c) = d(a \triangle c, 0)$	15,17
(19)	$d(a + c, ac) = d((a + c) \triangle ac, 0)$	18
(20)	$d(a + c, ac) = d(a \triangle c, 0)$	19
(21)	$d(a, c) = d(a + c, ac)$	18,20

Thus in a Boolean algebra the lattice diagonals are equal; it follows from (13) and (21) that the lattice quadrangles are rectangles. But this does not follow from (0) and (1) for arbitrary lattices, or even for Boolean lattices. For example, (0) and (1) hold in the four-element Boolean lattice that is illustrated in Figure 1. Here all the distances between distinct elements are equal to 1, except for $d(x + z, xz)$, which equals 2. This example allows us to show also that a strengthening of (1) that permits substitution in each argument of d is strictly stronger than (1) is, except in Boolean algebras. We certainly have

(22) $d(a, c) + d(\phi a \,\Delta\, \psi a, 0) \geq d(\phi c \,\Delta\, \psi c, 0)$ 1
(23) $d(a, c) + d(\phi a, \psi a) \geq d(\phi c, \psi c)$. 18,22

But (23) fails in the lattice illustrated, since

$$d(x, z) + d(x + x, xx) = 1 < 2 = d(x + z, xz).$$

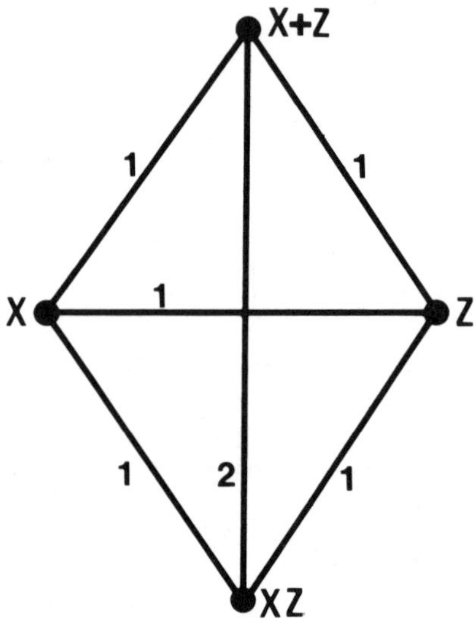

Figure 1

In fact, (21) follows from (0) and (23), for the latter yields in any lattice both

$$d(a, c) + d(a + a, aa) \geq d(a + c, ac)$$

and

$$d(a + c, ac) + d(a + c + a, a + c + c) \geq d(ac + a, ac + c) = d(a, c).$$

We shall explicitly use this fact in deriving (35) - (43) below.

The following isotony principles are also derivable from the assumption that d is defined simply on a lattice.

(24)	$d(c, a) \geq d(b + c, b + a)$	7
(25)	$c \geq b \geq a \rightarrow d(c, a) \geq d(c, b)$	24
(26)	$d(c, a) \geq d(bc, ba)$	8
(27)	$c \geq b \geq a \rightarrow d(c, a) \geq d(b, a)$	26

It is natural to wish to strengthen (25) and (27) to the additivity principle

(28) $c \geq b \geq a \rightarrow d(c, a) = d(c, b) + d(b, a)$.

Yet since (0) and (1) are satisfied by the trivial metric (which sets $d(a, b)$ equal to 1 except when $a = b$), it is obvious that, anyway when d is real-valued, (28) is a new principle that needs to be postulated separately. Note, however, that it readily yields (0), so that (1) and (28) can now be adopted as the axioms for partial identity (or diversity).

In a Boolean algebra we obtain for the complement:

(29)	$d(a, c) = d(a', c')$	6
(30)	$d(a, c) + d(a', c) = d(a \Delta c, 0) + d((a \Delta c'), 0)$	18
(31)	$d(a, c) + d(a', c) = d(1, a \Delta c) + d(a \Delta c, 0)$	30,29,3
(32)	$d(a, c) + d(a', c) = d(1, 0)$	31,28
(33)	$d(a, a') = d(a \Delta a', 0) = d(1, 0)$	18
(34)	$d(a, c) + d(a', c) = d(a, a')$	32,33

The following string of derivations will succeed in any lattice containing a function d for which (23) and (28), or even (21) and (28), hold, provided that the cancellation principle

if $x + y \geq z + y$ then $x \geq z$

holds in the ordered semigroup in which d takes its values. These derivations may become somewhat more comprehensible if continued reference is made to the lattice diagram in Figure 2.

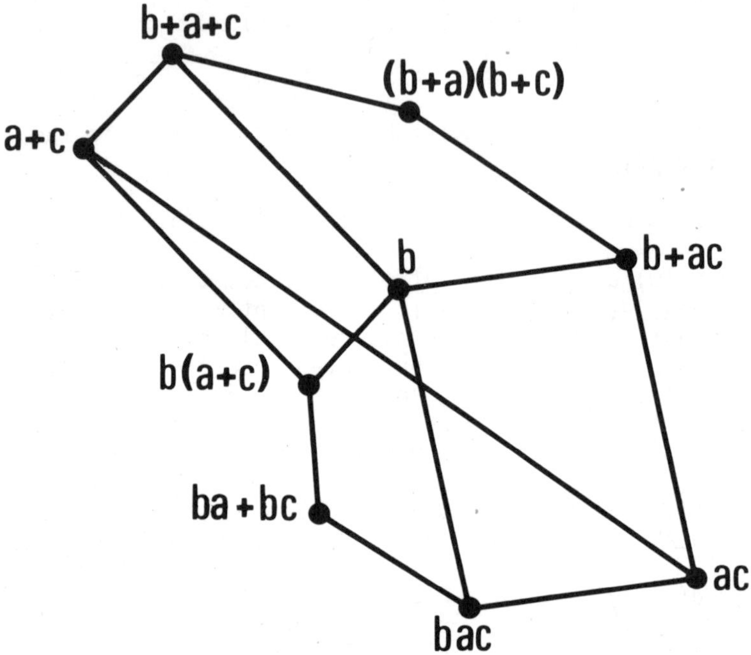

Figure 2

(35) $d(b + ac, ac) = d(b, bac)$
$d((b + a)(b + c), (bac + a)(bac + c))$ 13,6

(36) $d(b, bac) = d((b + a)(b + c), ac)$ 35,27

(37) $d(b + a + c, (b + a)(b + c)) + d(b, bac)$
$= d(b + a + c, ac)$ 28, 36

(38) $d(b + a, b + c) + d(b, bac) = d(b + a + c, ac)$ 37,21

Dually,

(39) $d(ba, bc) + d(b + a + c, b) = d(a + c, bac),$
3,6,13,21,25,28

whence

(40) $d(ba, bc) + d(b + a + c, b) = d(a + c, ac) +$
$+ d(ac, bac).$ 39,28

By addition of (38) and (40) we get

$$(41)\quad d(b + a, b + c) + d(ba, bc) + d(b + a + c, bac) = d(b + a + c, bac) + d(a + c, ac). \qquad 38,40,28$$

If we may now cancel $d(b + a + c, bac)$ from each side we obtain

$$(42)\quad d(b + a, b + c) + d(ba, bc) = d(a, c). \qquad 41,21$$

In a Boolean algebra no such additional cancellation principle needs to be assumed. For the Boolean identities

$$\begin{aligned} (b + a) \Delta (b + c) &= b'(a \Delta c) \\ &= (a \Delta c) \Delta b(a \Delta c) \\ ba \Delta bc &= b(a \Delta c) \end{aligned}$$

along with (18), used several times, and (28) allow us to transform one side of (42) into the other as follows:

$$\begin{aligned} d(b + a, b + c) &+ d(ba, bc) \\ &= d((a \Delta c) \Delta b(a \Delta c), 0) + d(b(a \Delta c), 0) \\ &= d(a \Delta c, b(a \Delta c)) + d(b(a \Delta c), 0) \\ &= d(a \Delta c, 0) = d(a, c) \end{aligned}$$

Thus (42) holds unrestrictedly in a Boolean algebra.

Since (13) and (27) both follow from (6) in any lattice, the next equation,

$$(43)\quad d(b + ac, ac) = d((b + a)(b + c), b + ac) + d(b + ac, ac) \qquad 28,35,36$$

holds in any lattice where (6) and (28) are satisfied. If the funtion d is real-valued, or otherwise permits cancellation, we may thus conclude

$$(44)\quad d((b + a)(b + c), b + ac) = 0.$$

If d is also a metric operation, so that a is identical with c whenever $d(a, c) = 0$, it follows that the underlying lattice must be distributive. Thus the existence in a lattice of a real-valued function satisfying (1) and (28) constrains the lattice to be a distributive one. It is of some interest that no more than modularity will be enforced if we replace (6) by the weaker restrictions (7) and (8). The inequality in (35), in particular, will no longer be derivable. This may be seen by setting d(a, c) equal to 1 in the non-distributive modular lattice of Figure 3 whenever a and c are adjacent in the ordering, and equal to 2 whenever they are not. Although (3), (4), (7), (8) and (28) all hold generally, (6) does not, since

$$\begin{aligned} d(y, yxz) &= 1 < 2 = d((y + x)(y + z), xz) \\ &= d((y + x)(y + z), (yxz + x)(yxz + z)). \end{aligned}$$

It is an easily established fact, however, that if d is a metric then (13) and (28), and therefore certainly (7), (8), and (28), are together enough to guarantee modularity. In

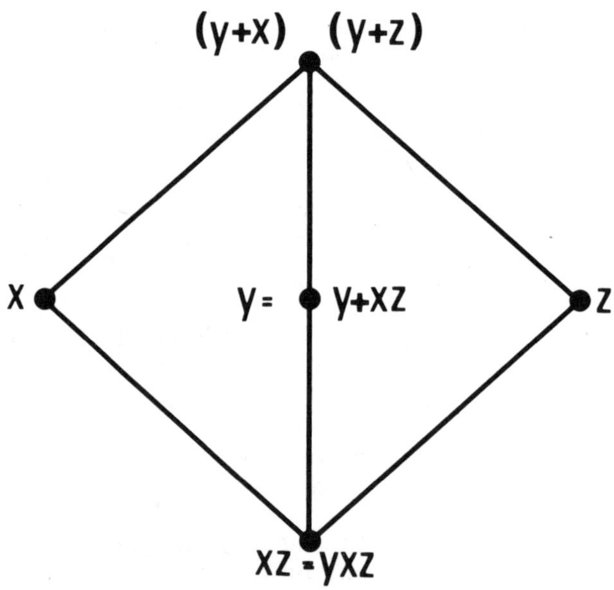

Figure 3

much the same way, the existence of a metric d satisfying (42) characterizes distributive lattices, whilst the weaker inequality

(43) $d(a, c) \geq d(b + a, b + c) + d(ba, bc)$

(from which (7), (8), and (28) may easily be derived) characterizes modular lattices.[2]

2. METRIC BOOLEAN ALGEBRAS

In this section we shall develop the numerical theory that is axiomatized by the two sentences (4) and (43) of the previous section, together with the assumption that the elements a, b, c, ... form a Boolean algebra. When supplemented with (34) this theory becomes virtually identical with one announced some years ago[3] as a numerical version of an autometric theory of Boolean algebra. We shall show, however, that (34) is itself a consequence of (4) and (43) in the numerical case, so may be dispensed with; indeed there is only one place where the succeeding derivations do not go through even if the range of d is assumed, as above, to be no more than a suitable ordered abelian semigroup with unit.

Sentences will be renumbered in this section, and for the sake of notational uniformity (4) is here relettered. Unless it is stated otherwise, annotations henceforth in this section will refer only to previously numbered sentences of the section.

Our axioms are

(0) $d(b, a) + d(b,c) \geq d(a, c)$
(1) $d(a, c) \geq d(ba, bc) + d(b + a, b + c)$.

The derivations proceed as follows.

(2)	$d(a, a) + d(a, a) \geq d(a, a)$	0
(3)	$d(a, a) \geq d(aa, aa) + d(a + a, a + a)$	1
(4)	$d(a, a) = 0$	2,3

It is at this point that some cancellation property for the semigroup, namely

$$\text{if } x + x = x \text{ then } x = 0,$$

needs to be assumed.

(5)	$d(b, a) \geq d(a, b)$	0,4
(6)	$d(b, a) = d(a, b)$	5
(7)	$d(a, c) \geq d(ba, bc)$	1
(8)	$d(a, c) \geq d(b + a, b + c)$	1
(9)	$d(a, c) \geq d(aca, acc)$	7
(10)	$d(a, c) \geq 0$	4,9
(11)	$d(a + c, c) \geq d(a(a + c), ac) = d(a, ac)$	7

(12)	$d(a, ac) \geq d(c + a, c + ac)$	8
	$= d(a + c, c)$	
(13)	$d(a + c, c) = d(a, ac)$	11,12

The additivity principle is also easily derived.

(14)	$c \geq b \geq a \rightarrow d(c, a) \geq d(b, a) + d(c, b)$	1
(15)	$c \geq b \geq a \rightarrow d(c, a) = d(c, b) + d(b, a)$	0,6,14
(16)	$d(a + c, ac) \geq d((a \Delta c)(a + c),$	
	$(a \Delta c) ac)$	7
(17)	$d(a + c, ac) \geq d(a \Delta c, 0)$	16
(18)	$d(a \Delta c, 0) \geq d(ac + (a \Delta c), ac + 0)$	8
(19)	$d(a \Delta c, 0) \geq d(a + c, ac)$	18
(20)	$d(a + c, ac) = d(a \Delta c, 0)$	17,19
(21)	$d(a, c) \geq d((a \Delta c)a, (a \Delta c)c)$	7
(22)	$d(a, c) \geq d(ac', a'c)$	21
(23)	$d(ac', a'c) \geq d(ac + ac', ac + a'c)$	8
(24)	$d(ac', a'c) \geq d(a, c)$	23
(25)	$d(a, c) = d(ac', a'c)$	22,24
(26)	$d(a, c) \geq d(a + a, a + c) + d(aa, ac)$	1
(27)	$d(a, c) \geq d(a + c, ac)$	0,26
(28)	$d(a + c, a) + d(a + c, c) \geq d(a, c)$	0
(29)	$d(a + c, a) + d(a, ac) \geq d(a, c)$	13,28
(30)	$d(a + c, ac) \geq d(a, c)$	15,29
(31)	$d(a, c) = d(a + c, ac)$	27,30
(32)	$d(a, c) = d(a \Delta c, 0)$	20,31
(33)	$d(a, c) = d(a', c')$	32

We may now derive

(34) $d(a, c) + d(a', c) = d(a, a')$

from (6), (32), (15), and (33) just as in the previous section we derived (34) from (3), (18), (28), and (29).

The following derivation demonstrates that the function $d(a, 0)$ is a valuation or dimension function.

(35)	$d(a + c, 0) = d(a + c, c) + d(c, 0)$	15
(36)	$d(a + c, 0) = d(a, ac) + d(c, 0)$	13,35
(37)	$d(a + c, 0) + d(ac, 0)$	
	$= d(a, ac) + d(ac, 0) + d(c, 0)$	36
(38)	$d(a + c, 0) + d(ac, 0)$	
	$= d(a, 0) + d(c, 0).$	15,37

Moreover, we have trivially

(39) $d(0, 0) = 0$ 4
(40) $d(a, 0) \geq 0.$ 10

By means of the abbreviation

$$m(a) = d(a, 0)$$

(38)-(40) may be cast in the form

(41) $m(0) = 0$
(42) $m(a) \geq 0$
(43) $m(a + c) + m(ac) = m(a) + m(c),$

showing that (0) and (1) do indeed yield the theory of unnormalized measures on a Boolean algebra.

3. UNNORMALIZED MEASURES

To complete our cycle of equivalences we shall show that the axioms for partial identity can be obtained from (41)-(43) above together with the definition

$$d(a, c) = m(a \Delta c).$$

For simplicity we shall retain throughout the metric rather than the measure notation. Our axioms for this section thus are

(0) $d(0, 0) = 0$
(1) $d(a, 0) \geq 0$
(2) $d(a + c, 0) + d(ac, 0) = d(a, 0) + d(c, 0)$
(3) $d(a, c) = d(a \Delta c, 0).$

We shall establish first the additivity principle, then the triangular inequality, and finally the generalization of the principle of the indiscernibility of identicals. The proofs are in general quite well known, and we shall not labour them unduly.

Note that

$$ca' = cb'a' + cba' = c(b + a)' + cba',$$

so that if $c \geq b \geq a$ then

$$ca' = cb' + ba'.$$

Moreover, if $y \geq x$ then $yx' = y \Delta x$, so we have

$$\text{if } c \geq b \geq a \text{ then } c \Delta a = (c \Delta b) + (b \Delta a).$$

We now apply (2), and use the fact that $cb'ba' = 0$ to obtain

$$d(c \Delta b, 0) + d(b \Delta a, 0) = d(c \Delta a, 0),$$

which in the presence of (3) yields the additivity principle

(4) $c \geq b \geq a \rightarrow d(c, a) = d(c, b) + d(b, a).$

To obtain triangularity, we note simply that

$$(b \Delta a) + (b \Delta c) \geq a \Delta c.$$

Since by (2) and (1)

$$d(b \Delta a, 0) + d(b \Delta c, 0) \geq d((b \Delta a) + (b \Delta c), 0),$$

we can apply (4) and then (3) to conclude that

(5) $d(b, a) + d(b, c) \geq d(a, c).$

Note further that, just as we prove the replacement theorem for sentential calculus,[4] we can prove that

$$a \leftrightarrow c \leq \phi a \leftrightarrow \phi c;$$

by contraposition of this inequality we obtain

$$a \Delta c \geq \phi a \Delta \phi c.$$

Applying (4), (3), and (1) we can conclude that

(6) $d(a, c) \geq d(\phi a, \phi c).$

Thus

(7) $d(a, c) + d(\phi a, b) \geq d(\phi a, \phi c) + d(\phi a, b),$
(8) $d(a, c) + d(\phi a, b) \geq d(\phi c, b)$

by (5). Finally we may conclude from (0) and (3) that

(9) d(a, a) = 0.

With (9) and (8) of this section we have returned to the axioms for partial identity with which we started.

4. CONCLUDING REMARKS

There is a considerable similarity between the ideas of this paper and some of the investigations of Katz,[5] who, however, proposes a many-valued logic of error. A proper comparison of our two approaches is urgently needed.

The drawbacks of the present theory, especially when proposed as a solution to the problem of verisimilitude (or distance from the truth), have been indicated elsewhere.[6] Suffice it to acknowledge here that the objections raised cannot be wholly disregarded, even if they (along with all other philosophical issues) have been disregarded in this paper. The present theory may be a valuable contribution to the problem of partial identity, or diversity, or dissimilarity; but if it is, that fact has not been demonstrated in the present paper.

Department of Philosophy,
University of Warwick

NOTES

1. This is a revised and corrected version of a paper that was contributed to the Second World Conference on Mathematics at the Service of Man, Las Palmas, Gran Canaria, 1982, and has appeared on pp. 507-9 of A. Ballester et al. (1982). An early report of these investigations was given in a lecture in April 1980 at the Istituto di Cibernetica del C.N.R. in Arco Felice (Napoli). I should like to take this opportunity of publicly thanking Dr Settimo Termini for his hospitality and generosity in arranging a number of visits to Arco Felice.
2. See Theorems 1, 7, and 5 respectively of Miller (1979). For closely related work see Theorems 12 and 13 of Monjardet (1981), and also Ganti (1978).
3. See the end of Miller (1977).
4. See for example section 26 of Kleene (1952).
5. See in particular Katz (1982).
6. See Miller (1978), section 4, and also section III of Miller (1983). A somewhat highhanded discussion is to be

found in section 1 of Oddie (1981); p.359 of Kuipers (1982) is perhaps too strong an antidote to this.

REFERENCES

Ballester, A., D.Cardus, and E.Trillas (eds): 1982, *Proceedings of the Second World Conference on Mathematics at the Service of Man*, Las Palmas.

Ganti, S.M.: 1978, 'Characterization of Pseudo-Metric Lattices', *Mathematics Seminar Notes*, University of Kobe 6, 341-4.

Katz, M.: 1982, 'Real-Valued Models with Metric Equality and Uniformly Continuous Predicates', *Journal of Symbolic Logic* 47, 772-92.

Kleene, S.C.: 1952, *Introduction to Metamathematics*, New York.

Kuipers, T.A.F.: 1982, 'Approaching Descriptive and Theoretical Truth', *Erkenntnis* 18, 343-78.

Miller, D.W.: 1977, 'New Axioms for Boolean Geometry', *Bulletin of the Section of Logic*, Institute of Philosophy and Sociology, Polish Academy of Sciences, Wroctaw 6, 53-63.

Miller, D.W.: 1978, 'On Distance from the Truth as a True Distance', in J. Hintikka, I. Niiniluoto, and E. Saarinen (eds), *Essays on Mathematical and Philosophical Logic*, Dordrecht, pp. 415-35.

Miller, D.W.: 1979, 'Metric Postulates for Modular, Distributive, and Boolean Lattices', *Bulletin of the Section of Logic*, Institute of Philosophy and Sociology, Polish Academy of Sciences, Wroctaw 8, 191-6.

Miller, D.W.: 1983, 'Truth, Truthlikeness, Approximate Truth', *Fundamenta Scientiae* (forthcoming).

Monjardet, B.: 1981, 'Metrics on Partially Ordered Sets - A Survey', *Discrete Mathematics* 35, 173-84.

Oddie, G.: 1981, 'Verisimilitude Reviewed', *British Journal for the Philosophy of Science* 32, 237-65.

S.V. Ovchinnikov

REPRESENTATIONS OF TRANSITIVE FUZZY RELATIONS

I. INTRODUCTION

Since Zadeh (I97I) introduced similarity relations and fuzzy orderings, a great deal of work has been done in studying properties of these relations and their applications. The reader is referred to recent volumes edited by Gupta and Sanchez (I982), Yager (I982), and Gupta, Ragade and Yager (I979) for a representative sample of publications in this area.

Fuzzy binary relation theory provides useful tools for dealing with vagueness and imprecision in many areas involving human judgment and perceptions. Social choice theory in economics and individual choice theories in economics and psychology are examples of applied areas where an approach based upon fuzzy set theory (Blin and Whinston I974, Nurmi I98I, Skala I982, Ovchinnikov I982a,b) might supply an effective alternative to the existing probabilistic theories.

This paper is not intended to develop a conceptual framework for representing vagueness in specific application areas, nor to add still another technique to the existing ones. In what follows, we will be mostly concerned with mathematical problems in studying the structure of fuzzy similarity and ordering relations. Moreover, our attention will be focused only on so-called max-product transitive relations. Being a specific case of transitivity, (see Goguen I967, Zadeh I97I), max-product transitivity, nevertheless, possesses properties making it possible to generalize the results of this paper. Recently, main theorems of the paper have been extended by L.Valverde and the author to a wide class of transitive relations covering almost all relations used in applications.

The representation theory developed in the paper is of relevance to hierarchical decision processes (Saaty I980), classification theory (Bezdek I98I, Ovchinnikov and Riera I982), choice theory (Ovchinnikov I982b) and related problem areas. We shall not discuss the possible applications of the theory to these areas in the present paper.

H. J. Skala, S. Termini, and E. Trillas (eds.), Aspects of Vagueness, 105–118.

2. TRANSITIVE FUZZY BINARY RELATIONS

A fuzzy binary relation R on a set U is defined as a fuzzy subset of the direct product U×U; this relation is completely determind by its membership function R(x,y) with domain UxU and range [0,I].

The most general transitivity property of fuzzy binary relation is defined by means of a binary operation star $*$ on [0,I] as follows

$$R(x,y) * R(y,z) \leq R(x,z) ,$$

for all $x,y,z \in U$. This property is called max-star transitivity, since it is based upon max-star composition (Zadeh I97I). A standard example of a star operation is the minimum operation defined by $a \wedge b = \min(a,b)$. Zadeh (I97I) suggested that max-product transitive relations defined by

$$R(x,y) \cdot R(y,z) \leq R(x,z) \tag{I}$$

might be useful in resolving certain difficulties in the theory of similarity relations. The structure of general max-star transitive relations waz analyzed by Ovchinnikov (I98I).

In order to obtain useful structural properties of max-star transitive relations, certain restrictions should be imposed on binary operation star. In most practically interesting cases this operation is a triangular norm (t-norm), i.e. a function

$$T : [0,I]^2 \rightarrow [0,I]$$

fulfilling the following properties

$T(x,I) = x$, (Boundary Condition)

$T(x,y) \leq T(u,v)$, provided $x \leq u$, $y \leq v$, (Monotonicity)

$T(x,y) = T(y,x)$, (Commutativity)

$T(T(x,y),z) = T(x,T(y,z))$ (Associativity) .

A t-norm defines a commutative semigroup on [0,I] with order-preserving operation. Minimum and product operations are examples of t-norms. For more details on t-norm theory the reader is refered to Klement (I98I) and Schweizer and Sklar (I963).

This paper is concerned with max-product transitive fuzzy binary relations. Moreover, two special classes of max-product transitive relations are separated in the following definition.

Definition I. A fuzzy binary relation R fulfilling properties:
i) $R(x,x) = I$, for all $x \in U$, and
ii) $R(x,y) \cdot R(y,z) \leq R(x,z)$, for all $x,y,z \in U$,
is said to be a fuzzy preordering (shortly, R-relation).
A fuzzy preordering S fulfilling property
$S(x,y) = S(y,x)$, for all $x,y \in U$,
is said to be a similarity relation (shortly, S-relation).

R-relations were studied by Ovchinnikov (I982b) in connection with choice theory. S-relations were introduced by Zadeh (I97I) in his theory of similarity relations. Karl Menger (I95I) suggested to study S-relations and a special case of R-relations in a general framework of 'probabilistic theory of relations'. In mathematical psychology binary choice probabilities very often satisfy max-product transitivity property (Luce and Suppes I965, Fishburn I973). S- and R-relations are of general interest in mathematics too, since negative logarithms of their membership functions are pseudo-distanses on U (nonsymmetric for R-relations). This connection provides interesting and fruitful interpretations of structural theorems for S- and R-relations.

3. STRUCTURAL THEOREMS

In this section we prove two theorems providing structural descriptions of R- and S-relations.

Theorem I. A fuzzy binary relation R is a fuzzy preordering if and only if there is a family F of positive functions on U such that

$$R(x,y) = \inf_{f \in F} \min \left\{ \frac{f(x)}{f(y)} , I \right\} \tag{2}$$

Proof. Let F be a family of positive functions on U. Then, for R defined by (2), we have

$$R(x,y) \cdot R(y,z) = \inf_{f \in F} \min \left\{ \frac{f(x)}{f(y)}, I \right\} \cdot \inf_{f \in F} \min \left\{ \frac{f(y)}{f(z)} , I \right\} \leq$$

$$\leq \inf_{f \in F} \min \left\{ \frac{f(x)}{f(y)} \cdot \frac{f(y)}{f(z)} , I \right\} = R(x,z),$$

and, obviously, $R(x,x) = I$, for all $x \in U$.

Let now R be a fuzzy preordering. We define a family of positive functions on U by

$$F = \left\{ f_{a,n},\ a \in U,\ n = 1,2,\dots \right\}$$

where

$$f_{a,n}(x) = \max\left\{ R(x,a),\ \frac{1}{n} \right\}.$$

Then, for any given $x,y \in U$,

$$\inf_{a,n} \min\left\{ \frac{f_{a,n}(x)}{f_{a,n}(y)},\ 1 \right\} \leq \inf_{n} \min\left\{ \frac{f_{y,n}(x)}{f_{y,n}(y)},\ 1 \right\} =$$

$$= \inf_{n} \min\left\{ f_{y,n}(x),\ 1 \right\} = R(x,y).$$

On the other hand, we have

$$R(x,y)\cdot f_{a,n}(y) = R(x,y)\cdot \max\left\{ R(y,a),\ \frac{1}{n} \right\} =$$

$$= \max\left\{ R(x,y)\cdot R(y,a),\ R(x,y)\cdot \frac{1}{n} \right\} \leq$$

$$\leq \max\left\{ R(x,a),\ \frac{1}{n} \right\} = f_{a,n}(x),$$

by max-product transitivity. Hence,

$$R(x,y) \leq \frac{f_{a,n}(x)}{f_{a,n}(y)},$$

which implies

$$R(x,y) \leq \inf_{a,n} \min\left\{ \frac{f_{a,n}(x)}{f_{a,n}(y)},\ 1 \right\}.$$

Therefore, (2) is proven. ∎

Theorem 2. A fuzzy binary relation S is a similarity relation if and only if there is a family F of positive functions on U such that

$$S(x,y) = \inf \min\left\{ \frac{f(x)}{f(y)},\ \frac{f(y)}{f(x)} \right\} \qquad (3)$$

The proof of this theorem is very similar to the previous one and is omitted.

Note, that theorems I and 2 say nothing regarding uniquness of representations (2) and (3). Actually, any given

R- or S-relation possesses various representations of this type. For instance, it follows from the proof that there are families of bounded functions providing representations (2) and (3).

The following corollary is a well-known result in topology.

Corollary. (Dugundji 1966, ch.XIII, theorem 5.2) Any metric space is isometrically embedded into a linear space of bounded functions with a supremum norm.

Proof. Let M be a metric space with a distance d(x,y). Then

$$S(x,y) = \exp(-d(x,y))$$

defines a similarity relation on M. By (3), we have

$$d(x,y) = \sup_{f \in F} \left| \ln f(x) - \ln f(y) \right| \tag{4}$$

for some family F of positive bounded functions on M. Note, that family F may be chosen such that all functions in F are bounded by the same constant. Then $x \mapsto \ln f(x)$ defines a mapping of the space M into the space of all bounded functions on F endowed by sup-norm. By (4), this mapping is isometry. ∎

This corollary provides a sort of geometric interpretation of representations (2) and (3).

4. DIMENSIONALITY

The following definition is based upon geometric interpretation of representation theorems.

Definition 2. A dimension of R- or S- relation is the minimal cardinal number of family F in (2) and (3).

Note, that the minimal cardinality under consideration exists, since any set of cardinal numbers is a well-ordered set.

One-dimensional R- and S- relations form important classes of fuzzy binary relations. By definition 2, any one-dimensional relation is defined by a single positive function f on U as follows

$$R(x,y) = \min\left\{\frac{f(x)}{f(y)}, 1\right\} \tag{5}$$

and

$$S(x,y) = \min\left\{\frac{f(x)}{f(y)}, \frac{f(y)}{f(x)}\right\}. \tag{6}$$

For a positive constant c, function cf defines the same fuzzy relations as f in (5) and (6). Hence, f may be regarded as a ratio-scale. This point of view is developed in section 6.

In view of definition 2, representation theorems I and 2 may be reformulated as follows.

Theorem 3. Any fuzzy preordering or similarity relation is an intersection of corresponding one-dimensional fuzzy relations.

In classical binary relation theory a similar statement holds. Namely, any preordering is represented as an intersection of weak orderings (a form of Szpilrajn I930 theorem). We will call one-dimensional R-relations fuzzy weak orderings; then theorem 3 may be regarded as an extension of Szpilrajn theorem.

5. CHARACTERIZATIONS OF ONE-DIMENSIONAL RELATIONS

One-dimensional fuzzy relations were defined in section 4 by means of representations (5) and (6). It is desirable to have an external description of the classes of fuzzy weak orderings and one-dimensional similarity relations, i.e. a description by means of certain characteristic properties. Various characterizations of one-dimensional fuzzy relations may be found in this section. One may regard these characterizations as a sort of axiomatization of the theories of fuzzy weak ordering and one-dimensional similarity relation. We begin with the following theorem.

Theorem 4. A reflexive fuzzy binary relation R is a fuzzy weak ordering if and only if it satisfies the following properties

i) $R(x,y) > 0$, for all $x,y \in U$,

ii) $\max\{R(x,y), R(y,x)\} = I$, for all $x,y \in U$,

iii) $R(x,y) = I$ and $R(y,z) = I$ imply together $R(x,z) = I$ and $R(z,x) = R(z,y)\cdot R(y,x)$, for all $x,y,z \in U$.

Proof. I) Necessity. Let R be a fuzzy weak ordering, i.e. it is represented by (5). Then i) and ii) are obviously, satisfied. Further, $R(x,y) = I$ and $R(y,z) = I$ imply together $f(x) \geq f(y) \geq f(z)$. Hence, $R(x,z) = I$, and

$$R(z,x) = \frac{f(z)}{f(x)} = \frac{f(z)}{f(y)} \cdot \frac{f(y)}{f(x)} = R(z,y) \cdot R(y,x).$$

2) Sufficiency. Let R be a reflexive fuzzy relation satisfying i)-iii). We define a positive function f on U by

$$f(x) = \begin{cases} R(x,a), & \text{if } R(a,x) = I, \\ \dfrac{I}{R(a,x)}, & \text{if } R(x,a) = I, \end{cases}$$

for some fixed $a \in U$. In order to prove (5), we consider the following cases:

a) $R(x,y) = R(y,a) = I$. Then, by iii), $R(x,a) = I$, and

$$\min\left\{\frac{f(x)}{f(y)}, I\right\} = \min\left\{\frac{R(a,y)}{R(a,x)}, I\right\} = \min\left\{\frac{I}{R(y,x)}, I\right\} = I.$$

b) $R(x,a) = R(a,y) = I$. Then, by iii), $R(x,y) = I$. We have

$$\min\left\{\frac{f(x)}{f(y)}, I\right\} = \min\left\{\frac{I}{R(a,x)\ R(y,a)}, I\right\} = I.$$

c) $R(y,x) = R(x,a) = I$. Then, by iii), $R(y,a) = I$, and $R(a,y) = R(a,x) \cdot R(x,y)$. We have

$$\min\left\{\frac{f(x)}{f(y)}, I\right\} = \min\left\{\frac{R(a,y)}{R(a,x)}, I\right\} = \min\left\{R(x,y), I\right\} = R(x,y).$$

d) $R(y,a) = R(a,x) = I$. Then, by iii), $R(y,x) = I$, and $R(x,y) = R(x,a) \cdot R(a,y)$. We have

$$\min\left\{\frac{f(x)}{f(y)}, I\right\} = \min\left\{R(x,a) \cdot R(a,y), I\right\} = \min\left\{R(x,y), I\right\} =$$

$= R(x,y)$.

e) $R(a,x) = R(x,y) = I$. Then, by iii), $R(a,y) = I$, and $R(y,a) = R(y,x) \cdot R(x,a)$. We have

$$\min\left\{\frac{f(x)}{f(y)}, I\right\} = \min\left\{\frac{R(x,a)}{R(y,a)}, I\right\} = \min\left\{\frac{I}{R(y,x)}, I\right\} = I.$$

f) $R(a,y) = R(y,x) = I$. Then, by iii), $R(a,x) = I$, and $R(x,a) = R(x,y) \cdot R(y,a)$. We have

$$\min\left\{\frac{f(x)}{f(y)}, I\right\} = \min\left\{\frac{R(x,a)}{R(y,a)}, I\right\} = \min\left\{R(x,y), I\right\} = R(x,y).$$

We proved (5), since all possible cases were considered. ∎

Note, that properties i) and ii) together might be regarded as a strong connectedness property; property iii) is a sort of transitivity property.

Theorems 5 - 7 provide various characterizations of one-dimensional similarity relations.

Theorem 5. A fuzzy binary relation S is a one-dimensional similarity relation if and only if there is a fuzzy weak ordering R such that

$$S(x,y) = \min\{R(x,y), R(y,x)\}.$$

Procf. Follows immediately from (5) and (6).

Note, that a similar statement is true for any similarity relation. Namely, S is a similarity relation if and only if there is a fuzzy preordering R such that

$$S = R \cap R^{-I}$$

where R^{-I} is an inverse fuzzy preordering.

Theorem 6. A fuzzy binary relation S is a one-dimensional similarity relation if and only if it satisfies the following properties:

i) $S(x,y) > 0$, for all $x,y \in U$,

ii) $S(x,y) = S(y,x)$, for all $x,y \in U$,

iii) there is a nonfuzzy weak ordering $\precsim$ on U such that

$$S(x,y) \cdot S(y,z) = S(x,z),$$

provided that $x \precsim y \precsim z$.

Proof. I) Necessity. Let S be a relation defined by (6). Obviously, i) and ii) are fulfilled. We define $\precsim$ on U by

$$x \precsim y \quad \text{iff} \quad f(x) \leq f(y).$$

Then, $\precsim$ is a weak ordering and iii) is satisfied, since

$$S(x,y) = \frac{f(x)}{f(y)}, \quad S(y,z) = \frac{f(y)}{f(z)}, \quad \text{and } S(x,z) = \frac{f(x)}{f(z)},$$

provided that $x \precsim y \precsim z$.

2) Sufficiency. Let S be a fuzzy binary relation with properties i)-iii). We define a positive function f on U by

$$f(x) = S(x,a), \text{ if } x \precsim a, \text{ and}$$

$$f(x) = \frac{I}{S(x,a)}, \text{ if } a \precsim x.$$

for some fixed $a \in U$. Note, that the function $f(x)$ is defined correctly, since $x \precsim a$ and $a \precsim x$ together imply $S(x,a) = I$, by i)-iii). In order to prove (6), let us consider the following cases:

a) $x \precsim y \precsim a$. Then

$$\min\left\{\frac{f(x)}{f(y)}, \frac{f(y)}{f(x)}\right\} = \min\left\{\frac{S(x,a)}{S(y,a)}, \frac{S(y,a)}{S(x,a)}\right\} =$$

$$= \min\left\{\frac{I}{S(x,y)}, S(x,y)\right\} = S(x,y), \text{ by iii).}$$

b) $x \precsim a \precsim y$. Then

$$\min\left\{\frac{f(x)}{f(y)}, \frac{f(y)}{f(x)}\right\} = \min\left\{S(x,a)\cdot S(a,y), \frac{I}{S(x,a)\cdot S(a,y)}\right\} =$$

$= S(x,y)$, by iii).

c) $a \precsim x \precsim y$. Then

$$\min\left\{\frac{f(x)}{f(y)}, \frac{f(y)}{f(x)}\right\} = \min\left\{\frac{S(y,a)}{S(x,a)}, \frac{S(x,a)}{S(y,a)}\right\} = S(x,y),$$

by iii).

By symmetry, these are all possible cases. ∎

Finally, we prove a theorem providing characteristic properties which do not involve any external structures like a nonfuzzy weak ordering $\precsim$ in theorem 6.

Theorem 7. A fuzzy binary relation S is a one-dimensional similarity relation if and only if it satisfies the following properties:

i) $S(x,y) > 0$, for all $x,y \in U$,

ii) $S(x,y) = S(y,x)$, for all $x,y \in U$,

iii) $\min\{S(x,y), S(y,z), S(x,z)\} =$
$= \max\{S(x,y)\cdot S(y,z), S(x,z)\cdot S(y,z), S(x,y)\cdot S(x,z)\}$,
for all $x,y,z \in U$,

iv) $S(x,y)\cdot S(y,z) = S(x,z)$, $S(y,z)\cdot S(z,u) = S(y,u)$, and $S(y,z) \neq I$ imply together $S(x,y)\cdot S(y,u) = S(x,u)$, for all $x,y,z,u \in U$.

Proof. We define an equivalence relation $\sim$ on U by

$$x \sim y \quad \text{iff} \quad S(x,y) = I.$$

Correctness of this definition follows immediately from i)-iii). This relation is a congruence relation for S. Indeed, let $x \sim y$, i.e. $S(x,y) = I$. Then, by iii), $S(x,z) = S(y,z)$, for all $z \in U$. Hence, a fuzzy binary relation

$$\tilde{S}(\tilde{x},\tilde{y}) = S(x,y),$$

$x \in \tilde{x}, y \in \tilde{y}$, $\tilde{x},\tilde{y} \in U/\sim$, is defined on a quotient-set $U/\sim$. We define a ternary relation B on $U/\sim$ by

$$B(\tilde{x},\tilde{y},\tilde{z}) \quad \text{iff} \quad S(x,y)\cdot S(y,z) = S(x,z).$$

The relation B satisfies the following properties:

I) $B(\tilde{x},\tilde{y},\tilde{x})$ implies $\tilde{x} = \tilde{y}$,

2) $B(\tilde{x},\tilde{y},\tilde{z})$ implies $B(\tilde{z},\tilde{y},\tilde{x})$,

3) $B(\tilde{x},\tilde{y},\tilde{z})$, $B(\tilde{y},\tilde{z},\tilde{u})$, and $\tilde{y} \neq \tilde{z}$ imply $B(\tilde{x},\tilde{y},\tilde{u})$,

4) either $B(\tilde{x},\tilde{y},\tilde{z})$, $B(\tilde{y},\tilde{z},\tilde{x})$ or $B(\tilde{z},\tilde{x},\tilde{y})$.

Altwegg (I950) proved that conditions I)-4) completely characterize betweenness relations in chains. Hence, there is a weak ordering $\lesssim$ on U, such that $\sim$ is an indifference relation for this weak ordering, and

$$S(x,y)\cdot S(y,z) = S(x,z) \quad \text{iff} \quad x \lesssim y \lesssim z \quad \text{or} \quad z \lesssim y \lesssim x.$$

This argument demonstrates that conditions of theorem 7 are equivalent to the conditions of theorem 6. ∎

6. FUZZY RELATIONAL SYSTEMS

General ideas of measurement theory are employed in this section; the reader is refered to Krantz et al.,(I97I) for the basic definitions and results of this theory.

Definition 3. A fuzzy relational system (f.r.s.)$\langle U,R\rangle$ is an ordered pair consisting of a set U and a fuzzy binary relation R on U. An f.r.s. $\langle U,R\rangle$ is a numerical fuzzy relational system (n.f.r.s.) if U is a set of real numbers or a subset in a real vector space.

Let F be a set; we denote by $(R^+)^F$ the set of all positive functions on F. Then

$$R(x,y) = \inf_{f \in F} \min\left\{ \frac{x(f)}{y(f)}, I \right\} \tag{7}$$

and

$$S(x,y) = \inf_{f \in F} \min \left\{ \frac{x(f)}{y(f)}, \frac{y(f)}{x(f)} \right\} \tag{8}$$

define fuzzy binary relations on $(R^+)^F$.

<u>Definition 4</u>. Numerical fuzzy relational systems N_o^F and N_s^F are defined by

$$N_o^F = \langle (R^+)^F, R \rangle, \text{ and}$$

$$N_s^F = \langle (R^+)^F, S \rangle,$$

where R and S are defined by (7) and (8), respectively.

As a special case, let us consider a singleton F. The corresponding n.f.r.s. are

$$N_o = \langle R^+, \min\left\{ \frac{x}{y}, I \right\} \rangle, \text{ and}$$

$$N_s = \langle R^+, \min\left\{ \frac{x}{y}, \frac{y}{x} \right\} \rangle$$

An important notion of a homomorphism of f.r.s. is introduced in definition 5.

<u>Definition 5.</u> Given any two f.r.s. $\langle U, R \rangle$ and $\langle V, Q \rangle$ a <u>homomorphism</u> from $\langle U, R \rangle$ into $\langle V, Q \rangle$ is defined as a mapping $f: U \to V$ such that

$$R(x,y) = Q(f(x), f(y))$$

for all $x,y \in U$.

Theorems I and 2 may be reformulated now as follows.

<u>Theorem 8</u>. A fuzzy binary relation R (resp. S) on U is a fuzzy preordering (resp. a similarity relation) if and only if there is a homomorphism from $\langle U, R \rangle$ (resp. $\langle V, Q \rangle$) into some n.f.r.s. N_o^F (resp. N_s^F).

Therefore, our structural theorems I and 2 may be also considered as representation results in a general framework of fuzzy relational systems.

We conclude this section with a theorem establishing representation and uniqueness results for one-dimensional n.f.r.s. N_o.

Theorem 9. Let $\langle U,R \rangle$ be an f.r.s. Then R is a fuzzy weak ordering if and only if there is a homomorphism f from $\langle U, R \rangle$ into N_o. Another function g defines a homomorphism if and only if there is a positive constant c such that $g(x) = c\, f(x)$, for all $x \in U$.

Proof. The representation part follows immediately from the definition of a fuzzy weak ordering (see (6)). To prove uniqueness, let us consider representations

$$R(x,y) = \min\left\{ \frac{f(x)}{f(y)}, I \right\} = \min\left\{ \frac{g(x)}{g(y)}, I \right\}, \text{ and}$$

$$R(y,x) = \min\left\{ \frac{f(y)}{f(x)}, I \right\} = \min\left\{ \frac{g(y)}{g(x)}, I \right\},$$

for all $x,y \in U$. They imply immediately

$$\frac{f(x)}{g(x)} = \frac{f(y)}{g(y)}, \text{ for all } x,y \in U. \blacksquare$$

7. CONCLUSION

A representation theory developed in this paper opens some interesting problems in mathematics and applied areas.

I) Let U be a topological space. It is easy to verify that fuzzy preorderings and similarity relations with continuous membership functions are represented by continuous functions in (2) and (3), respectively. The converse is not true: if F is a family of positive continuous functions on U, then, say, function S(x,y), given by (3), is not necessarily a continuous function on UxU. But it is certainly an upper semi-continuous function. A natural question is whether any upper semi-continuous similarity relation is represented by continuous functions in (3). The answer is affirmative if the space U is a locally compact space with a countable base. The proof is quite complicated and deserves a separate publication. A similar problem for fuzzy weak orderings is still open.

2) Let $\underline{S}$ be a set of all similarity relations on U. It is a poset with respect to the natural inclusion relation. Moreover, any subset of $\underline{S}$ has a greatest lower bound in $\underline{S}$. Hence, $\underline{S}$ is a complete lattice (Birkhoff I979, p.II2). Lattice $\underline{S}$ is a natural continuous analog of a partition lattice (Birkhoff I979, p.95). It is an interesting problem to study algebraic and topological properties of lattice $\underline{S}$,

for instance, a Galois connection given by theorem 2 between S and maximal cones of positive functions on U.

3) It was mentioned, already, in section 2 that binary choice probabilities sometimes have max-product transitivity property. A proper probabilistic interpretation of representation theorems would be of interest in, say, mathematical psychology and economics. There is a canonical way to transcribe properties described in terms of fuzzy set theory into a probabilistic framework, for any fuzzy set is a projection of a certain random set (see, for example, Goodman 1982). The author intends to investigate this problem in a future publication.

Department of Mathematics
San Francisco State University
San Francisco, California 94132, U.S.A.

ACKNOWLEDGEMENT

The author is grateful to Z.Domotor, E.Trillas and L.Zadeh for fruitful discussions.

REFERENCES

Altwegg,M.: 1950,'Zur axiomatik der teiweise geordneten mengen', Comm.Math.Helv. 24, 149-155.

Bezdek,J.C.: 1981, Pattern Recognition with Fuzzy Objective Function Algorithms, Plenum Press.

Birkhoff,G.: 1979, Lattice Theory, AMS Colloquium Publications, vol.25.

Blin,J.M.,and Whinston,A.B.: 1974,'Fuzzy sets and social choice', J.of Cybernetics 3, 28-36.

Dugundji,J.: 1966, Topology, Allyn and Bacon, Inc., Boston.

Fishburn,P.C.: 1973,'Binary choice probabilities: on the varieties of stochastic transitivity', J.of Math.Psych.10, 327-352.

Goguen,J.A.: 1967,'L-fuzzy sets', J.Math.Anal.and Appl. 18, 145-174.

Goodman,I.R.: 1982,'Fuzzy sets as equivalence classes of random sets' in Yager 1982.

Gupta,M.M.,and Sanchez,E.(eds): 1982, Fuzzy Information and Decision Processes, North-Holland.

Gupta,M.M.,Ragade,R.K.,and Yager,R.R.(eds): 1979, Advances in Fuzzy Set Theory and Applications, North-Holland.

Klement,E.P.: I98I,'Operations on fuzzy sets and fuzzy numbers related to triangular norms', in Proc.of the IIth Intern.Symp. on Multiple-Valued Logic, Norman, Oklahoma, U.S.A., 2I8-225.

Krantz,D.H.,Luce,R.D.,Suppes,P.,and Tversky,A.: I97I, Foundations of Measurement, vol.I, Academic Press, N.Y.

Luce,R.D.,and Suppes,P.: I965,'Preference, utility and subjective probability', in Handbook of Mathematical Psychology, vol.III,John Willey and Sons, Inc., 249-4I0.

Menger,K.: I95I,'Probabilistic theories of relations', Proc.Nat.Acad.Sci. 37, no.3, I78-I80.

Nurmi,H.: I98I,'Approaches to collective decision-making with fuzzy preference relations', Fuzzy Sets and Systems 6, 249-259.

Ovchinnikov,S.V.: I98I,'Structure of fuzzy binary relations' Fuzzy Sets and Systems 6, I69-I65.

Ovchinnikov,S.V.: I982a,'Social choice and Lukasiewicz logic' in Proc.of the I2th Int.Symp.on Multipl-Valued Logic, Paris, France, I63-I66.

Ovchinnikov,S.V.: I982b,'Choice theory for cardinal scales' in Gupta and Sanchez I982, 323-338.

Ovchinnikov,S.V.,and Riera,T.: I982,'On fuzzy classifications', in Yager I982, II9-I32.

Saaty,T.L.: I980, The Analytic Hierarchy Process, McGraw-Hill International.

Schweizer,B.,and Sklar,A.: I967,'Associative functions and abstract semigroups', Publ.Math.Debrecen I0, 69-8I.

Skala,H.J.: I982,'Modelling vagueness, in Gupta and Sanchez I982.

Szpilrajn,E.: I930,'Sur l'extension de l'ordre partiel', Fundamenta Mathematicae I6, 386-389.

Yager,R.R.(ed): I982, Recent Developments in Fuzzy Set and Possibility Theory, Pergamon Press.

Zadeh,L.A.: I97I,'Similarity relations and fuzzy orderings' Inf.Sci. 3, I77-200.

Aleš Pultr

FUZZINESS AND FUZZY EQUALITY

In the intuitive examples of fuzzy sets one often has more structure than just the degrees in which the elements belong. Typically, one encounters also various degrees of likeness of the elements with each other. Thus, e.g., in the "set" of all green objects in a given room we observe varying similarity of the coloring. In the "set" of very large natural numbers one has the obvious distance, in the "set" of patients diagnosed for a given illness one sees varying similarity of symptoms, etc. This is mostly connected with the fuzziness structure in the following way: if an element x is a member of X in some degree and another y is close (similar) to x in a high degree then y is a member in not much smaller degree than x (if, say, x is "very typically green" and the shade of y is very close to that of x, we will not much hesitate to classify y as green, too; if x is a very large natural number and y is close to x, we should take x for rather large; etc.).

The resulting fuzzy-equivalence (fuzzy-equality, tolerance) has been in that or other form considered before (see, e.g., [7],[8]), but it does not seem very popular. The reason for this may be in the natural wish to avoid unnecessary complexity. The aim of this paper is to show that the fuzzy-equivalence point of view can, in fact, make the calculating simpler. Roughly speaking, we show that fuzzy sets endowed with fuzzy-equivalences can be adequately represented by means of the fuzzy-equivalence alone. In the representation the fuzzy sets appear as couples (X', X'') of crisp subsets of universes with a kind of distance (resulting from the fuzzy-equalities) and one operates with them crisply (e.g., if fuzzy sets X_i are represented as (X'_i, X''_i), the union of the X_i in the usual sense gets represented as $(\bigcup X'_i, \bigcup X''_i)$, the product as $(\bigtimes X'_i, \bigtimes X''_i)$). Even the fuzzy power set of a fuzzy set is naturally expressed by means of a generalization of

H. J. Skala, S. Termini, and E. Trillas (eds.), Aspects of Vagueness, 119–135.

the metric in the Hausdorff hyperspace.

To avoid misunderstanding, let us make a few remarks on the distance functions issueing from fuzzy equivalences we will work with. Dealing with fuzzy notions it is natural to consider the value 0 in $x\varepsilon_o X$ as expressing poor belonging (if any), or, in $x=_o y$, a poor similarity of x and y, while the value 1 expresses in $x\varepsilon_1 X$ the (at least almost) full belonging, or in $x=_1 y$ the (at least almost) coincidence of the relevant features. Consequently, in the "metric" spaces we shall deal with it is handier to use, instead of the distance functions (with values diminishing when coming nearer), the nearness functions with values the higher the nearer one comes. The reader will certainly see that this is a technical preference only, and that the notion of L-nearness below has nothing to do with the well known continuity structure sometimes called nearness spaces ([5]).

The spaces with L-nearness are discussed in §§ 1 and 2. In § 3 we recall the usual operations with fuzzy sets and extend them in the obvious way to fuzzy sets with fuzzy-equalities. In § 4 the notion of L-nebula (which is the couple of subsets of a universe with an L-nearness mentioned above) is introduced. The main results are contained in § 5 where we show that fuzzy sets (with or without fuzzy equality) can be represented as nebulae (and vice versa) with reasonable behavior under operations.

§ 1. L-nearness and L-spaces with L a residuated lattice

1.1. A residuated lattice $L = L(\vee,\wedge,0,1,.,h)$ is a complete lattice endowed, moreover, with two further binary operations .,h such that

(i) L(.,1) is a commutative monoid,

(ii) . is non-decreasing in both variables,

(iii) h is non-increasing in the first variable and non-decreasing in the second one,

(iv) for all a,b,c $a.b \leqslant c$ iff $a \leqslant h(b,c)$.

1.2. The following identities are immediate consequences of the definition (with a particular role played by the condition (iv)):

(1) $a.\sup b_i = \sup(a.b_i)$,

(2) $h(a, \inf b_i) = \inf h(a,b_i)$

(3) $h(\sup a_i, b) = \inf h(a_i, b)$,

(4) $h(a.b,c) = h(a,h(b,c))$. □

1.3. Remarks: (1) Residuated lattices were, first, introduced in [1]: they appear in the literature also under the name of tensored lattices, or LC_∞-lattices ([3]). Viewing complete lattices as complete small thin categories, we see that the residuated ones coincide with the closed ones among them.

(2) In fact, h always exists and is uniquely determined if we have an $L(\vee,\wedge,0,1,.)$ satisfying (i) and 1.2(1). It is, however, more comfortable to work with h as explicitly given.

(3) Examples of residuated lattices:
(3.1) Complete Boolean algebras: put $.=\wedge$ and $h(a,b) = \bar{a}\vee b$.
(3.2) More generally, Heyting algebras are exactly the residuated lattices with $.=\wedge$.
(3.3) (Łucasiewicz lattice): L is the unit interval with minimum and maximum, $a.b = (a+b-1)\vee 0$, $h(b,c) = (1-b+c)\wedge 1$ (cf.[6]).
(3.4) $L(\vee,\wedge)$ as in (3.3), a.b the usual multiplication,

$h(b,c) = b^{-1}.c$ for $b>c$, 1 otherwise.

(4) It is easy to see that . is idempotent iff $.=\wedge$ (i.e., in the case of Heyting algebras). Otherwise there is always an a with $a.a<a$.

1.4. Here are a few of further basic properties of the operations in residuated lattices. The proofs are easy and are left to the reader as an exercise.
(1) $a.b\leq a\wedge b$,
(2) $a\leq b \Leftrightarrow h(a,b) = 1$,
(3) $h(a,b).h(b,c)\leq h(a,c)$,
(4) $h(1,a) = a$,
(5) $h(a,b) = h(a,b\wedge a)$.□

1.5. Let L be a residuated lattice. An L-nearness on a set P is a mapping

$$\nu: P\times P \to L$$

such that
(i) $\nu(x,x) = 1$

(ii) $\nu(x,y) = \nu(y,x)$,
(iii) $\nu(x,y).\nu(y,z) \leqslant \nu(x,z)$.

The system (P,ν) will be referred to as an L-space. We use the term subspace of (P,ν) in the obvious way, namely for the (Q,ν') with $Q \subset P$ and $\nu' = \nu | Q \times Q$.

1.6. Remark: One easily sees that the usual metric spaces are, in principle, obtained when taking for L the Łukasiewicz lattice (see 1.3(3.3)).

1.7. Observation and convention: The formula $\nu_o(a,b) = h(a,b) \wedge h(b,a)$ gives an L-nearness on L itself. Referring to L as a space we will always have in mind (L,ν_o).

Note that $\nu_o(a,1) = a$.

§ 2. Nearness of points and sets. Hausdorff superspace

2.1. For a subset $A \subset P$ we put

$$\nu(x,A) = \sup\{\nu(x,y) \mid y \in A\}.$$

2.2. Lemma: $\nu(x,A) \geqslant \nu(x,B).\inf\{\nu(y,A) \mid y \in B\}$.

Proof: For each y we have (recall 1.2.(1))

$$\nu(x,A) \geqslant \sup_{z \in A}(\nu(x,y).\nu(y,z)) = \nu(x,y).\sup_{z \in A}\nu(y,z) = \nu(x,y).\nu(y,A).$$

Put $d = \inf\{\nu(y,A) \mid y \in B\}$. We obtain $\nu(x,A) \geqslant \nu(x,y).d$ for any $y \in B$ and hence finally

$$\nu(x,A) \geqslant \sup_{y \in B}(\nu(x,y).d) = (\sup_{y \in B}\nu(x,y)).d = \nu(x,B).d. \ \square$$

2.3. Lemma: $\nu(x,\cup A_i) = \sup_i \nu(x,A_i)$.

Proof: Obviously $\nu(x,A_j) \leqslant \nu(x,\cup A_i)$ and hence $\sup \nu(x,A_i) \leqslant \nu(x,\cup A_i)$. On the other hand, if $y \in \cup A_i$, we have, for some j, $\nu(x,y) \leqslant \nu(x,A_j) \leqslant \sup \nu(x,A_i)$ so that also $\nu(x,\cup A_i) \leqslant \sup \nu(x,A_i)$. $\square$

2.4. For $A, B \subset (P, \nu)$ put

$$\tilde{\nu}(A,B) = \inf_{x\in A} \nu(x,B).\inf_{y\in B} \nu(y,A).$$

2.5. Proposition: $\tilde{\nu}$ is an L-nearness on the set of all subsets of (P,ν).

Proof: Obviously $\tilde{\nu}(A,A) = 1$ and $\tilde{\nu}(A,B) = \tilde{\nu}(B,A)$. Now, we have (by 1.1. (ii) and 2.2.)

$$\tilde{\nu}(A,B).\tilde{\nu}(B,C) = \inf_{x\in A} \nu(x,B).\inf_{y\in B} \nu(y,A).\inf_{z\in B} \nu(z,C).$$

$$.\inf_{u\in C} \nu(u,B) = \inf_{x\in A} \nu(x,B).\inf_{z\in B} \nu(z,C).\inf_{u\in C} \nu(u,B).\inf_{y\in B} \nu(y,A) \leqslant$$

$$\leqslant \inf_{x\in A} (\nu(x,B).\inf_{z\in B} \nu(z,C)).\inf_{u\in C}(\nu(u,B).\inf_{y\in B} \nu(y,A)) \leqslant$$

$$\leqslant \inf_{x\in A} \nu(x,C).\inf_{u\in C} \nu(u,A) = \tilde{\nu}(A,C). \square$$

2.6. The set of all subsets of (P,ν) endowed with the nearness $\tilde{\nu}$ will be called the Hausdorff superspace of (P,ν) and denoted by

$$\exp(P,\nu).$$

One sees that this is an obvious generalization of the well-known synonymous construction for metric spaces ([4]).

2.7. Remark: (P,ν) is naturally embedded into $\exp(P,\nu)$ by representing $x\in P$ as $\{x\}$. Indeed, we have $\tilde{\nu}(\{x\},\{y\}) = \nu(x,y)$.

2.8. Remarks: Define a ν-closure of a set A as

$$\bar{A} = \{x \mid \nu(x,A) = 1\},$$

and call A closed if $\bar{A} = A$.

Using 2.2. one proves easily that

$$\nu(x,\bar{A}) = \nu(x,A)$$

and in consequence $\bar{\bar{A}} = \bar{A}$.

The ν-closure is easily shown to be additive iff $(a,b\in L \,\&\, \&\, a\vee b = 1) \Rightarrow (a=1 \text{ or } b=1)$.

The definition in 1.5. can be modified by replacing (i) by stronger

(i*) $\nu(x,y) = 1$ iff $x = y$

to obtain a notion closer to metric spaces (as it is, it corresponds rather to the pseudometric ones). All what is said in this article can be easily modified for this stronger structure, the basic change being in replacing "subsets" in the definition in 2.6 by "closed subsets".

§ 3. Fuzzy sets and fuzzy equality

3.1. Let us recall the basic definition of fuzzy sets and operations with them (cf., e.g., [10],[2]). L is a residuated lattice.

An L-fuzzy set is a mapping

$$X: |X| \to L$$

where $|X|$ is a set (the carrier of X). We write (with obvious interpretation)

$$x \,\varepsilon_a X \text{ for } X(x) \geq a.$$

The union, intersection, product and tensor product of fuzzy sets are given by the formulas

$$|\bigcup_{i\varepsilon J} X_i| = \bigcup_{i\varepsilon J} |X_i|, \ (\bigcup_{i\varepsilon J} X_i)(x) = \sup\{X_i(x) \mid i \text{ such that } x\varepsilon |X_i|\},$$

$$|\bigcap_{i\varepsilon J} X_i| = \bigcap_{i\varepsilon J} |X_i|, (\bigcap_{i\varepsilon J} X_i)(x) = \inf\{X_i(x) \mid i\varepsilon J\},$$

$$|\prod_{i\varepsilon J} X_i| = \prod_{i\varepsilon J} |X_i|, \ (\prod_{i\varepsilon J} X_i)((x_i)_J) = \inf\{X_i(x_i) \mid i\varepsilon J\},$$

$$|X \otimes Y| = |X| \times |Y|, \ (X \otimes Y)(x,y) = X(x).Y(y).$$

3.2. ("Mappings up to degree a" and fuzzy exponentitation.) If X,Y are L-fuzzy sets, we write

$$f:_a X \to Y \text{ if } f: |X| \to |Y| \text{ and } (x\varepsilon_b X \Rightarrow f(x)\,\varepsilon_{ab} Y).$$

We write $X \overset{w}{\subset} Y$ ("the weak inclusion"), if $|X| \subset |Y|$, and further

$$X \subset_a Y \text{ if } X \overset{w}{\subset} Y \text{ and } (x\varepsilon_b X \Rightarrow x\varepsilon_{ab} Y).$$

(Cf. [2],[9]).

Remark: Obviously, $X \subset_a Y$ iff $j:_a X \to Y$ for the inclusion mapping $j: |X| \subset |Y|$. On the other hand, if we define Γ, the graph of a mapping φ, by

$$|\Gamma| = \{(x,\varphi(x))\,|\,x\varepsilon|X|\}, \qquad \Gamma(x,\varphi(x)) = X(x),$$

we have $\Gamma \subset_a X \times Y$ iff $\varphi :_a X \to Y$.

3.3. The everyday-life examples of fuzzy sets often have, in addition, a structure of "imperfect equality" $=_a$ which we may reasonably assume to satisfy the implication

(1) $x =_a y$ & $y =_b z \Rightarrow x =_{ab} z$.

Moreover, it is usually (intuitively) such that if an element x is almost equal to y and y is in X in some degree, then x is in X in not much worse degree. This second matter will be formalized in 3.5 below.So far note that if we define

$$\nu(x,y) \geqslant a \text{ iff } x =_a y,$$

we obtain an L-nearness (the condition (1) represents the 1.5.(iii)).
This leads to the following definition:

An <u>L-fuzzy set with L-equality</u> (briefly, an <u>L-fe set</u>) (X,ν) consists of an L-fuzzy set $X\colon |X| \to L$ and a L-space $(|X|,\nu)$.

3.4. We write

$$(X,\nu) \overset{w}{\subset} (Y,\mu)$$

if $(|X|,\nu)$ is a subspace of $(|Y|,\mu)$. The basic operations with L-fe sets are defined as follows:

If $(X_i,\nu_i) \overset{w}{\subset} (X,\nu)$, we put

$$\bigcup_{J(X,\nu)} (X_i,\nu_i) = (\bigcup_J X_i, \nu \backslash |\bigcup_J X_i|),$$

$$\bigcap_J (X_i,\nu_i) = (\bigcap_J X_i, \nu \,|\, |\bigcap_J X_i|).$$

For a general system $((X_i,\nu_i))_{i\varepsilon J}$ put

$$\bigtimes_J (X_i,\nu_i) = (\bigtimes_J X_i, \mu) \text{ where } \mu((x_i)_J,(y_i)_J) =$$

$$= \inf \nu_i(x_i,y_i).$$

Finally put

$(X_1,\nu_1) \otimes (X_2,\nu_2) = (X_1 \otimes X_2,\nu)$ where $\nu((x_1,x_2),(y_1,y_2)) =$

$$= \nu_1(x_1,y_1) \cdot \nu_2(x_2,y_2).$$

Remark: Obviously, the usual L-fuzzy sets can be viewed as L-fe sets with trivial (discrete) equalities. Then, the operations coincide with those from 3.1.

3.5. An L-fe set (X,ν) is said to be saturated if

(*) $x \,\varepsilon_a X \;\&\; y =_b x \Rightarrow y \,\varepsilon_{ab} X.$

This is an obvious formalization of the second property from 3.3. Handier, although perhaps less lucid, is the following evident refomulation of (*):

(**) $\forall x \forall y \;\; \nu(x,y) \cdot X(y) \leq X(x).$

3.6. Proposition: (X,ν) is saturated iff X viewed as a mapping between L-spaces satisfies the inequality $\nu_o(X(x),X(y)) \geqslant \nu(x,y).$

Proof: By 1.1.(iv), (**) is, further, equivalent with

$$\forall x \forall y \;\; \nu(x,y) \leq h(X(y),X(x)). \quad \square$$

3.7. For an L-fuzzy set put

$K(X) = \{(x,a) \mid x \varepsilon |X|,\; a \leqslant X(x)\}.$

3.8. Let (X,ν) be an L-fe set. Define

$$\exp(X,\nu) = (Y,\mu),$$

where

$$|y| = \{Z \mid Z \text{ saturated}, |Z| \subset |X|\}.$$

$$Z \,\varepsilon_a Y \text{ iff } Z \subset_a X,$$

and

$$\mu(Z_1,Z_2) = \tilde{\nu}(K(Z_1),K(Z_2)) \text{ where } \nu \text{ is defined on}$$

$|X| \times L$ by $\nu((x_1,x_2),(y_1,y_2)) = \nu(x_1,y_1) \cdot \nu_o(x_2,y_2).$

3.9. Proposition: $\exp(X,\nu)(Z) = \inf_{x \varepsilon Z} h(Z(x), X(x)).$

Proof: We have $(\exp X)(Z) \geqslant a$ iff $\forall x (Z(x) \geqslant b \Rightarrow X(x) \geqslant ab)$

iff $\forall x (X(x) \geqslant a.Z(x))$ iff $\forall x (h(Z(x),X(x))) \geqslant a$ iff

iff $\inf h(Z(x),X(x)) \geqslant a.$ $\square$

3.10. Remarks: (1) Realize that the definition of the L-nearness μ in the power L-fe set in 3.8. is quite natural. Consider the "distance" of L-fe subsets with coinciding carriers.

(2) There arises a natural question as to whether the L-fe sets obtained from the saturated ones by operations are again saturated. It is not hard to prove the affirmative directly. We do not do it here, however, since it will follow free from the sequel.

§ 4. L-nebulae

4.1. An L-nebula is a system $N = ((P,\nu);K,R)$ where (P,ν) is an L-space and K,R are subsets of P. The set K will be referred to as the kernel and the set R as the range of the nebula.

4.2. Since the notion of an L-nebula and a representation of L-fuzzy sets and L-fe sets as nebulae is the main item of this article, let us discuss here briefly the intuitive background. The space (P,ν) is a universe in which we consider the elements we are interested in. Its size is not all that important. It could be replaced by a (P',ν') of which (P,ν) is a subspace, or by a subspace of (P,ν) large enough to contain R, and the situation described by the nebula would be virtually the same. The range corresponds to the carrier of a fuzzy set. The kernel is a system of typical elements with the property we try to describe (elements which have the property undisputedly); it is not necessarily a subset of the range. Let us give a few examples:

(a) The "set" of very large natural numbers: Take for the universe the set of natural plus ∞, with the nearness, say, $\nu(m,n) = 1 - m^{-1}n^{-1}.|m-n|$, $\nu(n,\infty) = 1 - n^{-1}$. For R take the set of natural numbers, $K = \{\infty\}$.

(b) The "set" of all yellow objects in a given area: Choose a nearness function expressing tolerance in interchangeability of colors. R is the set of objects which can be taken for yellow considered tolerantly (although someone could take some of them for, say, ochre, light brown, or greenish). K is a set of undisputedly yellow ones; may be there is none in the area considered - then choose a really yellow sample elsewhere.

(c) The "set" of patients suffering a particular disease. There are border cases with not very typical symptoms, but still included in R, while K contains only the typical ones, perhaps also imaginary textbook examples.

4.3. The <u>product</u> $\underset{J}{\times}(P_i,\nu_i)$ of a system $((P_i,\nu_i))_{i\varepsilon J}$ of L-spaces is defined as $(\underset{J}{\times}P_i,\nu)$ with $\nu((x_i)_J,(y_i)_J) =$
$= \inf \nu_i(x_i,y_i)$.

Further, we define the <u>tensor product</u> $(P_1,\nu_1)\otimes(P_2,\nu_2)$ of two L-spaces as $(P_1\times P_2,\nu)$ with $\nu(x_1,x_2),(y_1,y_2)) =$
$= \nu_1(x_1,y_1).\nu_2(x_2,y_2)$.

<u>Remarks</u>: Thus, the products (tensor products) of L-fe sets in 3.4. are carried by the products (tensor products) of the carrying spaces. Also observe that the $K(Z_i)$ in 3.8 are considered as elements of $\exp((|X|,\nu)\otimes L)$.

The product is the categorial product in the naturally defined category of L-spaces (where the morphisms correspond to non-expanding mappings of metric spaces). Also, the tensor product is one in this category, having a natural right adjoint.

4.4. We write $N = ((P,\nu);K,R) \overset{w}{\subset} ((P',\nu');K',R') = N'$ if (P,ν) is a subspace of (P',ν') and $R\subset R'$. If, moreover, also $K\subset K'$, we write $N\subset N'$. The former case is referred to as <u>weak inclusion</u>, the latter one simply as <u>inclusion</u> of nebulae.

The operations are defined as follows:

If $N_i = ((P_i,\nu_i);K_i,R_i) \overset{w}{\subset} N = ((P,\nu);K,R) \ (i\varepsilon J)$,

$$\bigcup_J{}_N N_i = ((P,\nu);\ \bigcup_J K_i,\ \bigcup_J R_i).$$

$$\bigcap_J N_i = ((\bigcap_J P_i,\nu \mid \bigcap_J P_i);\ \bigcap_J K_i,\ \bigcap_J R_i).$$

For an arbitrary system $(N_i)_{i\in J}$ of nebulae put

$$\bigtimes_J N_i = (\bigtimes_J (P_i,\nu_i);\ \bigtimes_J K_i,\ \bigtimes_J R_i).$$

Further, put

$$N_1 \otimes N_2 = ((P_1,\nu_1)\otimes(P_2,\nu_2);\ K_1\times K_2,\ R_1\times R_2).$$

4.5. For $N = ((P,\nu);\ K,R)$ put
$\exp N = (\exp(P,\nu)\times \exp(P,\nu);\ \exp K\times \exp K,\ \exp R\times \exp R)$.
Thus, the kernel of exp N consists of all the nebulae included in N, the range consists of those weakly included there. (Here, of course, the subnebulae are represented simply by their kernels and ranges, the universe being silently understood as the original one.)

§ 5. L-nebulae as L-fe sets and vice versa. An L-nebula can be viewed as an L-fe-set in a very natural way (see 5.1.). In this section we will show that, on the other hand, any saturated L-fe set can be represented by a nebula and, moreover, there is a representation such that the operations with the L-fe sets get represented as the corresponding operations with nebulae. Thus the (more complicated, since involving counting in L) L-fe set operations reduce, in principle, to operating with couples of crisp sets (see 5.14. and 5.15. below).

5.1. For a nebula $N = ((P,\nu);\ K,R,)$ define an L-fe set $\hat{\mathfrak{F}}(N)$ by putting
$\hat{\mathfrak{F}}(N) = (\mathfrak{F}(N),\nu|R)$ where $|\mathfrak{F}(N)| = R$, $\mathfrak{F}(N)(x) = \nu(x,K)$.
We say that N realizes an L-fe set (X,ν) if $(X,\nu) = \hat{\mathfrak{F}}(N)$.

5.2. Propostion: Each $\hat{\mathfrak{F}}(N)$ is saturated.

Proof: We have $\nu(x,y)\ .\ \nu(y,K) \leq \nu(x,K)$ by 2. □

5.3. Convention: In spaces $P\otimes L$ we replace the points $(x,1)$ by x. Thus, P is embedded into $P\otimes L$ as a subspace instead of the $P\times\{1\}$ level.

5.4. For a saturated L-fe set (x,ν) put

$$\mathcal{N}(X,\nu) = ((|X|,\nu)\otimes L;\ K(X),|X|)$$

(K(X) from 3.7).

5.5. <u>Lemma</u>: In a saturated (X,ν) one has
$\nu(x,y).h(a,X(y)) \leqslant h(a,X(x))$.

Proof: We have $X(y).\nu(x,y) \leqslant X(x)$, hence $X(y) \leqslant h(\nu(x,y), X(x))$. Hence further, by 1.1.(iii), 1.2.(4) and the commutativity of the multiplication, $h(a,X(y)) \leqslant h(a,h(\nu(x,y), X(x)) = h(\nu(x,y),h(a,X(x))$ and by 1.2.(4) again $h(a,X(y)) . \nu(x,y) \leqslant h(a,X(x))$. □

5.6. Proposition: Let (X,ν) be saturated. Then one has in $(|X| \times L,\bar{\nu}) = (|X|.\nu) \otimes L$ the equality
$$\bar{\nu}((x,a).K(X)) = h(a,X(x)).$$

Proof: Consider a $(y,b) \varepsilon K(X)$. We have $b \leqslant K(y)$ and hence, by 5.5
$$\bar{\nu}((x,a),(y,b)) = \nu(x,y).\nu_o(a,b) \leqslant \nu(x,y).h(a,b) \leqslant$$
$$\leqslant \nu(x,y).h(a,X(y)) \leqslant h(a,X(x)).$$
Thus, $\bar{\nu}((x,a),K(X)) \leqslant h(a,X(x))$. On the other hand, $(x,X(x)\wedge a)$ is in $K(X)$ and we have by 1.4.(5)
$$\bar{\nu}((x,a),(x,X(x)\wedge a)) = \nu_o(a,X(x)\wedge a) = h(a,X(x)\wedge a) = h(a,X(x)). \quad \square$$

5.7. <u>Theorem</u>: If (X,ν) is saturated, then
$\hat{\mathfrak{F}}\mathcal{N}(X,\nu) = (X,\nu)$.

Thus, each saturated L-fe set is realized by an L-nebula.

Proof: (Recall Convention 5.3.) By 5.6. and 1.4.(3),
$\bar{\nu}(x,K(X)) = h(1,X(x)) = X(x)$. □

5.8. <u>Remark</u>: From the proof of 5.6. one sees that $X(x) = \bar{\nu}(x,K(X)) = \bar{\nu}(x,(x,X(x)))$. □

5.9. The following is evident

<u>Proposition</u>: $(X,\nu) \overset{w}{\subset} (Y,\mu) \Rightarrow \mathcal{N}(X,\nu) \overset{w}{\subset} \mathcal{N}(Y,\mu)$.
$$N \overset{w}{\subset} M \Rightarrow \hat{\mathfrak{F}}(N) \overset{w}{\subset} \hat{\mathfrak{F}}(M). \quad \square$$

5.10. <u>Proposition</u>: $\mathcal{N}(\bigcap_J (X_i,\nu_i)) = \bigcap_J \mathcal{N}(X_i,\nu_i)$.

Proof: The coincidence of the universa is obvious. Further, we have

$(x,a) \varepsilon \bigcap K(X_i)$ iff $\forall i\ X_i(x) \geqslant a$ iff $a \leqslant \inf X_i(x) = (\bigcap X_i)(x)$

iff $(x,a) \varepsilon K(\bigcap(X_i,\nu_i))$. □

5.11. Remark: One does not have, in general, $\mathcal{N}(\bigcup(X_i,\nu_i)) = \bigcup \mathcal{N}(X_i,\nu_i)$. In case of a linearly ordered L, however, this equality holds at least for finite J.

5.12. Proposition: (1) $\hat{\mathfrak{F}}(\bigcup_{JN} N_i) = \bigcup_{J\,\hat{\mathfrak{F}}(N)} \hat{\mathfrak{F}}(N_i)$.

(2) $\hat{\mathfrak{F}}(N_1 \otimes N_2) = \hat{\mathfrak{F}}(N_1) \otimes \hat{\mathfrak{F}}(N_2)$

Proof: The coincidence of the carrying universa is evident. Further, we have by 2.3.

$\hat{\mathfrak{F}}(\bigcup N_i)(x) = \nu(x, \bigcup K_i) = \sup \nu(x,K_i)$

which yields (1). Finally, we have (use 1.2.(1))

$\hat{\mathfrak{F}}(N_1 \otimes N_2)(x_1,x_2) = \sup \{\nu_1(x_1,y_1).\nu_2(x_2,y_2) \mid (y_1,y_2) \varepsilon$

$\varepsilon K_1 \times K_2\} = \sup\limits_{y_1 \varepsilon K_1} (\sup\limits_{y_2 \varepsilon K_2} (\nu_1(x_1,y_1).\nu_2(x_2,y_2)) =$

$= \sup\limits_{y_1 \varepsilon K_1} (\nu_1(x_1,y_1). \sup\limits_{y_2 \varepsilon K_2} \nu_2(x_2,y_2)) =$

$= (\sup\limits_{y_1 \varepsilon K_1} \nu_1(x_1,y_1)).(\sup\limits_{y_2 \varepsilon K_2} \nu_2(x_2,y_2)) = \hat{\mathfrak{F}}(N_1)(x_1).\hat{\mathfrak{F}}(N_2)(x_2)$

which yields (2). □

5.13. Lemma: Let $N_i = ((P_i,\nu_i); K_i,R_i)$ be such that for each $x \varepsilon R_i$ there is an $\bar{x} \varepsilon K_i$ such that $\nu_i(x,K_i) = \nu_i(x,\bar{x})$. Then

$$\hat{\mathfrak{F}}(\mathop{\times}_J N_i) = \mathop{\times}_J \hat{\mathfrak{F}}(N_i).$$

Proof: Again, the coincidence of the carriers is evident. Denote by ν the nearness in $\times P_i$. We have for $x = (x_i) \in \times R_i$,

$$a = \hat{\mathcal{F}}(\times N_i)(x) = \nu(x, \times K_i) = \sup\{\nu((x_i),(y_i)) \mid (y_i) \in \times K_i\} = \sup\{\inf_{i\in J} \nu_i(x_i,y_i) \mid (y_i) \in \times K_i\},$$

$$b = (\times \hat{\mathcal{F}}(N_i))(x) = \inf \nu_i(x_i,K_i) = \inf_{i\in J} \sup\{\nu_i(x_i,y_i) \mid y_i \in K_i\}$$

and consequently $a \leqslant b$.

On the other hand, consider $\bar{x} = (\bar{x}_i)$. We have

$$\sup\{\nu_i(x_i,y_i) \mid y_i \in K_i\} = \nu_i(x_i,\bar{x}_i)$$

and hence

$$\inf_{i\in J} \nu_i(x_i,\bar{x}_i) \leqslant a \leqslant b = \inf_{i\in J} \nu_i(x_i,\bar{x}_i). \quad \square$$

5.14. Theorem: For $(X_i,\nu_i) \overset{w}{\subset} (X,\nu)$ we have

(1) $\hat{\mathcal{F}}(\bigcap_{i\in J} \mathcal{N}(X_i,\nu_i)) = \bigcap_{i\in J} (X_i,\nu_i)$,

(2) $\hat{\mathcal{F}}(\bigcup_{i\in J\ \mathcal{N}(X,\nu)} \mathcal{N}(X_i,\nu_i)) = \bigcup_{i\in J (X,\nu)} (X_i,\nu_i)$.

Generally

(3) $\hat{\mathcal{F}}(\times_{i\in J} \mathcal{N}(X_i,\nu_i)) = \times_{i\in J} (X_i,\nu_i)$.

(4) $\hat{\mathcal{F}}(\mathcal{N}(X_1,\nu_1) \otimes \mathcal{N}(X_2,\nu_2)) = (X_1,\nu_1) \otimes (X_2,\nu_2)$.

Proof: The expressions on the left hand side in (1) and (2) make sense by 5.9.

(1): $\bigcap(X_1,\nu_1) = \hat{\mathcal{F}}\mathcal{N}(\bigcap(X_i,\nu_i)) = \hat{\mathcal{F}}(\bigcap \mathcal{N}(X_i,\nu_i))$

by 5.7. and 5.10.

(2): $\hat{\mathfrak{F}}(\cup\mathcal{N}(X_i,\nu_i)) = \cup\hat{\mathfrak{F}}\mathcal{N}(X_i,\nu_i) = \cup(X_i,\nu_i)$

by 5.12.(1) and 5.7.

(3) $\hat{\mathfrak{F}}(\mathsf{X}\,\mathcal{N}(X_i,\nu_i)) = \mathsf{X}\,\hat{\mathfrak{F}}\mathcal{N}(X_i,\nu_i) = \mathsf{X}(X_i,\nu_i)$

by 5.8.,5.13.and 5.7.

(4) $\hat{\mathfrak{F}}(\mathcal{N}(X_1,\nu_1)\otimes\mathcal{N}(X_2,\nu_2)) = \hat{\mathfrak{F}}\mathcal{N}(X_1,\nu_1)\otimes$

$\otimes\hat{\mathfrak{F}}\mathcal{N}(X_2,\nu_2) = (X_1,\nu_1)\otimes(X_2,\nu_2)$ by 5.12.and 5.7.

5.15. Before formulating a statement on exponentiation analogous to the statements concerning the other operations, let us realize that it cannot be of the form $\hat{\mathfrak{F}}(\exp\mathcal{N}(X,\nu)) = \exp(X,\nu)$. The carrier of $\hat{\mathfrak{F}}(\exp\mathcal{N}(X,\nu))$ is a set of L-nebulae, that of $\exp(X,\nu)$ is a set of L-fe sets. Thus, one has to have a translation of fe-subsets to subnebulae. Such a translation really exists and, moreover, it is coherent with the construction $\mathcal{N}$. We have

<u>Theorem</u>: Define a mapping

$$\mathcal{N}_1 : |\exp(X,\nu)| \to R(\exp\mathcal{N}(X,\nu))$$

by putting $\mathcal{N}_1(Y,\nu') = (K(Y),|Y|)$ for $(Y,\nu') \overset{w}{\subset} (X,\nu)$. Then

(1) $\mathcal{N}(Y,\nu') = ((|Y|,\nu');\ \mathcal{N}_1(Y,\nu))$,

(2) $\mathcal{N}_1$ preserves the L-nearness,and

(3) $\hat{\mathfrak{F}}(\exp\mathcal{N}(X,\nu))\circ\mathcal{N}_1 = \exp(X,\nu)$.

<u>Proof</u>: (1) is obvious.
(2) Let us denote the L-nearness in $(|X|,\nu)\otimes L$ by ν_1.

Recall that the L-nearness is $|\exp(X,\nu)|$ (see 4.8.)is given by

$$\mu(Y_1,Y_2) = \tilde{\nu}_1(K(Y_1),\ K(Y_2)).$$

Now, for the nearness μ'in $R(\exp\mathcal{N}(X,\nu))$ we also have (recall 4.5.

$$\mu'(\mathcal{N}_1(Y_1),\ \mathcal{N}_1(Y_2)) = \tilde{\nu}_1(K(Y_1),K(Y_2))\wedge\tilde{\nu}_1(|Y_1|,|Y_2|) =$$

$$= \tilde{\nu}_1(K(Y_1),K(Y_2)),$$

since obviously $\tilde{\nu}_1(K(Y_1),(Y_2)) \leq \tilde{\nu}_1(|Y_1|,|Y_2|)$.

(3) In the product of two spaces we have $\nu((u,v), U\times V) = \nu(u,U)$ whenever $v \in V$. Thus,

$$\hat{\mathcal{F}}(\exp\mathcal{N}(X,\nu))(\mathcal{N}_1(Y,\nu')) = \nu_1(K(Y), \exp K(X)).$$

Take an $M \subseteq K(X)$. We have

$\tilde{\nu}_1(K(Y),M), = \inf\{\nu_1((x,a),M) \mid (x,a)\in K(Y)\}. \inf\{\nu_1((y,b), K(Y)) \mid (y,b)\in M\} \leq \inf\{\nu_1((x,a),M) \mid (x,a)\in K(Y)\} \leq$

$\leq \inf\{\nu_1((x,a), K(X)) \mid (x,a)\in K(Y)\}$.

Thus, by 5.6. and 3.9.

$$\tilde{\nu}_1(K(Y),M) \leq \inf\{h(a,X(x)) \mid a \leq Y(x)\} = \inf\{h(Y(y),X(x)) \mid x\in |Y|\} = \exp(X,\nu)(Y).$$

On the other hand, we have by 6.10., 6.6. and 3.9.

$$\tilde{\nu}_1(K(Y),K(X)\cap K(Y)) = \tilde{\nu}_1(K(Y),K(X\cap Y)) =$$

$$= \inf\{\nu_1((x,a),K(X\cap Y)) \mid (x,a)\in K(Y)\} =$$

$$= \inf\{h(a,X(x)\wedge Y(x)) \mid (x,a)\in K(Y)\} =$$

$$= \inf_x h(Y(x),X(x)\wedge Y(x)) = \inf_x h(Y(x),(X(x)) =$$

$$= \exp(X,\nu)(Y).$$

Thus, $\hat{\mathcal{F}}(\exp\mathcal{N}(X,\nu))(\mathcal{N}_1(Y)) = \exp(X,\nu)(Y)$. □

Matematicko-fyzikálini fakulta, Universita Karlova,
Sokolovská 83, 18600 Praha 8, Czechoslovakia

NOTE

Reprinted from Comment. Math. Univ. Carolinae 23, 2, (1982).

REFERENCES

[1] Dillworth, R.P., and M. Ward: 1939, 'Residuated lattices', Trans. Amer. Math. Soc. 45, 335-354.

[2] Goguen, J.A.: 1967,'L-Fuzzy sets', J. Math. Anal. Appl. 18, 145-174.

[3] Goguen, J.A.: 1968/69, 'The logic of inexact concepts', Synthese 19, 325-373.

[4] Hausdorff, F.: 1914, 'Grundzüge der Mengenlehre', Leipzig.

[5] Herrlich, H.: 1974, 'Topological structures', Math. Centre Trancts 52, 59-122.

[6] Łucasiewicz, J.: 1970, 'Selected works', Studies in Logic and Foundations of Mathematics, North-Holland Publ. Comp., Amsterdam, and PWN, Warszawa.

[7] Pavelka, J.: 1977, 'On L-fuzzy semantics', Weiterbildungszentrum für Kybernetik u. Rechentechnik, TU Dresden, Heft 21, 12-25.

[8] Pavelka, J.: 1979, 'On Fuzzy Logic I, II, III', Zeitschrift f. math. Logic u. Grundl. d. Math. 25, 45-52.

[9] Pultr, A.: 1976, 'Closed Categories of L-Fuhhy Sets', Weiterbildungszentrum f. math. Kybernetik u. Rechentechnik, TU Dresden, Heft 16, 60-68.

[10] Zadeh, L.A.: 1965, 'Fuzzy sets', Information and Control 8, 338-353.

Heinz J. Skala

LARGE SOCIETIES AND INDIVIDUAL STRATEGY SELECTION: A CASE STUDY OF AMBIGUITY

SUMMARY. In a recent paper Dubey and Shapley (1979) discuss a difficulty which occurs if we assume that the agents of a non-atomic game are completely free with respect to their individual strategic decisions and the gross events in the model - at - large are supposed to result from these decisions. It has been noted by Shapley that the principal source of difficulty lies in the conflict between the non-cooperative ideal of strictly independent decisionmaking and the technical demands of integrability.

As it is not completely evident how the appropriate formulation of the problem should look like we discuss various possible procedures which avoid the mentioned difficulty. It is argued that for phenomena in which "large" collections are involved we often have no clear cut intuition. Thus at various stages we have to make decisions both with respect to the formulation of the problem and the proposed solution. The problem of Dubey and Shapley serves as case study for this.

After a short review of Shapley's procedure we discuss other procedures involving the concept of a referee. The observation that neither the use of the full axiom of choice nor the σ-additivity of the measure space can be motivated by purely economic arguments leads to alternative settings avoiding the technical difficulty. Finally the problem is discussed from the point of view of measure extensions.

1. INTRODUCTION

The present paper is an extended version of an invited paper presented at the "second world conference on mathematics at the service of man". By means of a recent problem from mathematical economics we discuss various levels of ambiguity which we encounter when we look for mathematical representations of actual world phenomena.

(1) Do we accept, as Cantor did, the fiction of actually infinite sets or do we want to deal with the phenomenon

H. J. Skala, S. Termini, and E. Trillas (eds.), Aspects of Vagueness, 137–160.

of infinity in accordance with our everyday experience, that is to say, as a phenomenon which is involved in the observation of large, incomprehensible sets as Vopěnka (1979) proposed. In his set theory all sets are formally finite, but large sets have subcollections which behave as infinite sets do in classical set theory. One should note that in Vopěnka's Alternative Set Theory (AST) vagueness is not abandoned. Actually he argues that vagueness is just the essence of "infinity". In contrast to the theory of fuzzy sets no intermediate truth values are involved one just has to distinguish between vague and precise concepts. We shall not say more about AST as in this Volume the reader can find Sochor's excellent survey article about recent developments in AST. In conclusion we only remark that there is an intimate connection between AST and nonstandard methods. In fact, particular non-well-founded models are used in order to show the consistency of AST with respect to classical set theory. We shall not make much use of nonstandard methods but only remark that it would be quite natural to start with a *-finite representation of our problem and then obtain a measurable representation via the Loeb space approach (see Loeb (1975)). We shall come back to this approach in a subsequent paper.

Related to AST is also Mycielski's (1981) (recursive axiomatizable) theory FIN in which actual infinity is also abandoned. It seems that FIN is sufficient to formalize all "applicable" mathematics.

(2) Even if we decide to base our investigations on classical set theory, as will be done in this paper, it is sometimes not clear which of the many different models of set theory or "possible universes of sets" should serve as carrier for our economic problem.

For the following let us denote by ZF the usual axioms of Zermelo-Fraenkel set theory. ZFC denotes ZF together with the axiom of choice (AC).

Usually one is not concerned with the existence of different models of set theory as the investigations are considered to take place in the "true" universe of sets V, i.e., the part of Cantor's universe of sets which can be axiomatized in ZF. For most problems this is surely a sound practice as, when viewed from inside different models have the same usual properties. For example inside each model of ZFC there exists a unique field of real numbers of cardinality $2^{\aleph_0}$, however, when viewed from outside the corresponding fields of real numbers need not be isomorphic. It is commonly assumed that the properties which are not decided by ZFC and by which various models of set theory differ are "esoteric". For the user of mathematics esoteric can only mean that the property in question has no impact on his problem. We shall see, however, that for our problem and others (see e.g. Skala (1981)) some "esoteric" statements play a crucial role. Let us state for example a few well-known "esoteric" statements not decided by ZFC:

(α) The Continuum Hypothesis (CH).

(β) Gödel's axiom "V=L" which just says that all sets are constructible thus yielding a very poor universe of sets.

(γ) Every PCA-set (= projection of a coanalytic set) is Lebesgue measurable.

(δ) Martin's Axiom (MA).

In the context of mathematical economics the truth of (γ) would imply that Aumann's (1967) measurable utility theorem holds without any connectedness assumption. Let us also mention the following well-known relationships between the statements (α)-(δ): (β) implies (α), (β) implies not (γ), (δ) plus not CH implies (γ).

As we have just observed that statements which are not decided in ZFC may well have an impact on economic theory we have two options. Either we accept only such statements which can be decided in ZFC (or any other theory); all other statements are neither true nor false. Such an attitude would perhaps lead to too few truths. Or we make decisions (i.e., accept hypotheses) depending on our interests. In a certain sense such an attitude will lead to "soft" results. It is clear that our decisions should not be completely arbitrary though they need not be deducted from "self evident" principles. They may well be accepted because they systematically explain our area of interest. In this paper we shall consider two sorts of hypotheses which have been very successful, large cardinal hypotheses and various determinateness hypotheses. We would like to note that large cardinal hypotheses may be looked upon as extrapolations of principles already inherent in the axioms of ZFC (see Reinhardt (1974)). For a penetrating discussion of the relativity of set theoretic notions the philosophically minded reader is referred to Putnam's (1980) "Models and Reality".

(3) Even if we single out a class of intended models, we are faced with some type of ambiguity. For example, for the solution of our problem it will be important to decide whether the individual strategy spaces are real intervals or consist only of rational numbers (after all what should it mean to send π entities of a good to the market). The decision to use a certain measure space is another example to the point.

We shall observe in the following that the various levels of ambiguity are intimately connected. E.g., the decision to work in Solovay's (1970) model allows us to demand the strongest (and perhaps most natural) version of independent individual decisionmaking.

2. NON-ATOMIC GAMES IN STRATEGIC FORM - THE PROCEDURE OF SHAPLEY

We shall be interested in the following in large non-cooperative games. As measure theory provides us with natural tools for modelling economic problems where many agents are involved we shall follow here this approach. There is also another methodology which uses tools from nonstandard analysis. The relationship between both approaches is now quite clear since one can convert non-standard to standard measure spaces (see Loeb (1975)). The converse was already known much earlier.

In particular, if we want that neither a single agent nor a certain preformed coalition should have an influence, then a non-atomic measure space is an appropriate model. It was Aumann (1964) who for the first time used this idea in order to prove Edgeworth's conjecture.

In a recent paper Dubey and Shapley (1979) discussed non-cooperative game models of price formation and trade in pure exchange economies. They proved, for example, that in a certain model one can establish an equivalence between the non-cooperative (or strategic) Nash equilibria and the classical competitive ones provided that certain requirements are met concerning the monetary instruments, non-satiation of agents, desirability of goods and adequacy of the penalities. We shall not go into the details of the models presented by Dubey and Shapley (1979) but only note that the ideal of strictly independent individual decisionmaking causes a serious technical and methodological difficulity. This difficulity simply occurs because the function resulting from the individual decisions usually will be non-measurable. As Shapley and Dubey remark: "It goes to the heart of a basic distinction that divides the game-theoretical methodology and the more behavioristic approach: Is the economic agent a free decisionmaker or an automation?"

In the first part of their paper Dubey and Shapley simply ignore this problem and take the equilibrium solutions to be measurable by definition. In the second part Shapley proposes a procedure by which the problem may be attacked.

After shortly reviewing the procedure of Shapley we shall discuss various other approaches to the problem. The results are stated in the most simple form; some of them will appear elsewhere in a more general setting. No proofs are given in detail, however, the hints given should be sufficient for filling up the gaps.

Let $(A,\mathbb{A},p)$ be a non-atomic measure space, where A denotes the set of agents, $\mathbb{A}$ is the σ-algebra of measurable coalitions, and p is a non-atomic σ-additive measure. (If not otherwise stated we take A to be the closed unit interval with the Borel sets and Lebesgue measure.)

In order to keep things simple we assume that the agents' strategy spaces are all the same, say $S_x := [0,1]$ for all $x \in A$. As the agents act completely independent any function $g : A \to [0,1]$ may occur as strategy selection. The difficulty arises when the declared intentions of the agents, which are represented by g, are to be implemented in the market as a whole. This requires to pass from g to a set function $G : \mathbb{A} \to \mathbb{R}$. The most obvious way to do so is by direct integration:

$$G(B) = \int_B g dp.$$

This, however, leaves us with the non-measurable functions which will be the typical result of non-cooperative actions and which cannot be abandoned by economic arguments. The problem is how to handle non-measurable strategy selections without affecting measurable Nash equilibria.

Shapley's original idea was to trap g between measurable functions and to define

$$G_*(B) = \max \{\int_B f dp : f \leq g \text{ a.e., } f \text{ p-measurable}\},$$

$$G^*(B) = \min \{\int_B f dp : f \geq g \text{ a.e., } f \text{ p-measurable}\}.$$

Obviously, one may choose any G between G_* and G^* and leave the decision to the modeller. For example the modeller may choose any real number $\alpha \in [0,1]$ and define

$$G = \alpha G_* + (1-\alpha) G^*.$$

As, without any further motivation, this is hardly a satisfactory procedure Shapley proposes a heuristically acceptable method. His basic idea is to demand some sort of cooperation for "small" (in the sense of the measure p) coalitions and put the job of insuring integrability into the hand of the agents.

In our simplified model we may for example assume that p(B) denotes the total amount of some good available to the coalition B. Let us imagine that there is a referee to whom the "small" coalitions have to transmit their intentions. The resulting set function q need not be additive. Under weak assumptions, however, q has an additive part q_o which closely approximates q on small sets and which may be implemented.

In order to give an example of a possible type of joint behaviour which leads to a set function with additive part we assume that a coalition acts in accordance with its most timid member, i.e.

$$q(B) = p(B) \inf g(B).$$

It will not be a surprise that in this case one gets $g_o = G_*$. In his paper Shapley also considers more sophisticated group decision rules.

We would like to note that there is a strong type of joint behaviour which leads to a measurable strategy selection allowing us to proceed by direct integration. Suppose that a referee chooses some $\varepsilon > 0$ and announces a finite measurable partition $\Pi = \{B_1, \ldots, B_n\}$ of A such that $p(B_i) \leq \varepsilon$. The members of each B_i are expected to agree upon a strategy to which every agent in B_i adheres. The resulting function

$$g(x) = \alpha_i, \text{ if } x \in B_i$$

is a simple function and thus measurable.

One might suspect that it is possible to demand some type of group decision rule only for a certain partition of

A into sets of measure zero in order to get a measurable function g. However it is an immediate consequence of a quite deep theorem of Solovay (see Prikry (1977)) that this is not the case.

Theorem 1. Let $\{B_i : i \in I\}$ be any partition of $[0,1]$ into pairwise disjoint sets with Lebesgue measure zero. Then there is an $I' \subset I$ such that $\bigcup_{i \in I'} B_i$ is not Lebesgue measurable.

In this context it is interesting to observe that one may start with a nonstandard measure space and impose the type of joint behaviour just discussed on (infinitely) small coalitions in order to obtain (standard) measurable strategy selections. More precisely assume that the set of agents is modelled by the nonstandard interval $*[0,1]$. Let $\omega \in *\mathbb{N}-\mathbb{N}$ be an infinite natural number, let $\mathbb{A}$ be the internal algebra generated by the (infinitely small) intervals $\{[0,1/\omega), [1/\omega),[2/\omega), \ldots, [(\omega-1)/\omega,1]\}$ and let p be the *-finitely additive measure on $\mathbb{A}$ such that $p([i/\omega, (i+1)/\omega)) = 1/\omega$. Assume that we demand cooperation on the part of the coalitions $[i/\omega, (i+1)/\omega)$ such that the resulting strategy selection $g : *[0,1] \rightarrow *[0,1]$ is constant on each such interval. g is measurable with respect to $\mathbb{A}$. Hence, by Loeb's (1975) argument the standard part of g, st(g), is a standard real valued function which is measurable with respect to the smallest standard σ-algebra generated by $\mathbb{A}$. Moreover, as st(g) takes only finite values the integrals $\int_B g dp$ and $\int_B st(g) d\bar{p}$ differ only by an infinitesimal for any $B \in \mathbb{A}$. Here $\bar{p}$ denotes the measure on the smallest standard σ-algebra which contains $\mathbb{A}$ and is generated by the standard part st(p) of the measure p on $\mathbb{A}$.

3. OTHER PROCEDURES INVOLVING THE CONCEPT OF A REFEREE

Instead of looking for acceptable procedures involving some kind of cooperative action which anyway distroys the non-cooperative ideal we may imagine a referee announcing some restriction as to the strategy space of the agents. The single agents are then expected to declare their intentions according to the announced restriction without any cooperation whatsoever. If the resulting strategy selection g is not measurable it is up to the referee to choose a measurable function g' without damaging too many individual declarations.

Case 1: The referee announces a countable set from which the agents have to choose.

Theorem 2. Let $(A,\mathbb{A},p)$ be a probability space and let A' be any subset of A. If the function $g:A'\to\mathbb{R}$ assumes at most countably many different values on A', then there exists a measurable function g' and a subset $A''\subset A'$ such that $p^*(A'') = p^*(A')$ and $g(x) = g'(x)$ for all $x \in A''$.
(p^* denotes the outer measure induced by p.)

In particular, if the referee receives declarations form all $x \in A$, then he can find a measurable g' which agrees with the received declarations on a set of outer measure one. Theorem 2 is due to Saks and Sierpinski (1927) and has been used by them to derive theorem 3.

Case 2: The referee announces an $\varepsilon > 0$. The agents in turn choose arbitrary ε-intervals from which the referee can select in order to obtain a measurable function.

Theorem 3. Let $(A,\mathbb{A},p)$ be a probability space. Let g be any set-valued function such that for all x $g(x)$ is an ε-interval in $\mathbb{R}$. Then there exists a measurable function g' and an $A'\subset A$, $p^*(A') = 1$, such that $g'(x) \in g(x)$ for all $x \in A'$.

Again, the referee can manage to satisfy the intentions of all agents save for a set of inner measure zero.

Using a result of Łoś and Marczewski (1949) we can find a measure extension $(A,\overline{\mathbb{A}},\overline{p})$, such that in Theorem 3 we have $\overline{p}(A') = 1$. A similar remark holds for Theorem 2.

If we do not assume that Continuum Hypothesis (CH), then by using Martin's Axiom (MA) we can generalize Theorem 2 for functions g assuming less than $2^{\aleph_0}$ different values. As we shall refer to MA on several occasions let us state here its topological version (see e.g. Kunen (1980)): For any $\kappa > \aleph_0$ MA(κ) is the following assertion: If X is any compact Hausdorff space such that there is no uncountable family of pairwise disjoint nonempty open subsets of X and G_α are dense open sets for $\alpha < \kappa$, then $\bigcap_{\alpha<\kappa} G_\alpha \neq \emptyset$.

MA states that MA(κ) is true for every $\kappa < 2^{\aleph_0}$.

Let us note that MA($\aleph_0$) is a theorem of ZFC and that MA(κ) implies $\kappa < 2^{\aleph_0}$.

In view of Theorem 3 one is tempted to suppose that measurable selection theorems (for a recent survey consult Wagner (1977)) should be very useful for our purpose. However, in order to cope with the non-cooperative ideal we need a measurable selection theorem in which only a reasonable assumption on the values of the set-valued function is used. However, in all selection theorems known to the author there occurs some sort of hypothesis on the measurability and/or on the graph of the set-valued function which can be satisfied in our problem only at the cost of introducing some type of cooperative behaviour.

In order to give the reader some feeling how intricate the problem is we shall consider a situation where every agent independently may choose a non-void set of $\mathbb{R}$. Let us assume for the following that the continuum hypothesis (CH) does not hold, i.e. $2^{\mathbb{N}} > \aleph_1$. Martin and Solovay (1970) have proved the following

<u>Theorem 4.</u> It is consistent with ZFC that every subset of real numbers of cardinality $\aleph_1$ is a coanalytic set (briefly CA-or Π^1_1-set).

As in absence of CH the continuum may be very large we may wish to model our society by a set of cardinality $\aleph_1$ only. Let $A \subset [0,1]$ be any set of cardinality $\aleph_1$ and let the agents in A choose non-void subsets of $\mathbb{R}$. Denoting the resulting set-valued function by g and using the axiom of choice we can always obtain a function $g' : A \to \mathbb{R}$ such that $g'(X) \in g(x)$ for all $x \in A$. As $g'^{-1}(B)$ is at most of cardinality $\aleph_1$, for every Borel-set $B \subset \mathbb{R}$ we may by Theorem 4 consistently assume that g' is universally measurable. (It is a classical theorem of measure theory that all analytic sets and hence coanalytic sets are universally measurable.)

However, in view of Martin's Axiom it is also consistent with ZFC that A must have measure zero. Even more generally it can be shown that the following holds:

<u>Theorem 5.</u> It is consistent relative to ZFC that there exists no non-atomic probability space $(A, \mathbb{A}, m)$ such that $|A| < 2^{\aleph_0}$.

As it is well known that $ZFC + MA + 2^{\aleph_0} > \aleph_1$ is consistent we only have to show that $MA + 2^{\aleph_0} > \aleph_1$ implies that no such probability space exists. This, however, can be performed by using a recent result of Koumoullis (1981). Another proof may be found in Skala (1982).

Remark. The trouble with selection theorems is that even assuming the agents to choose nice sets they do not produce automatically a set valued function with a nice graph if they are too many i.e. have the cardinality $2^{\mathbb{N}}$. For example, if the agents choose Borel sets, then it follows from Theorem 4 that the resulting graph can consistently be assumed to be a Σ_2^1 - or PCA-set (= projection of a coanalytic set) if only $\aleph_1 < 2^{\mathbb{N}}$ agents are involved. Obviously, for a continuum of agents this does not hold. As we shall see later for Σ_2^1-sets there exist nice selections even in absence of the axiom of choice.

Without relying on a large cardinal assumption it is not known to the author whether one can consistently assume: "There exists a set $A \subset \mathbb{R}$ of Lebesgue measure > 0 such that every set-valued function g on A which takes values in the non-void Borel subsets of $\mathbb{R}$ (or any other reasonable value hypothesis) admits a measurable selector."

With respect to the classical selection theorem it is quite clear that the weaker the assumptions on the graph of set valued function are the weaker are the demands on the cooperative behaviour in order to produce such a set valued function. However, it seems to be quite hard to obtain an operational description of this demands.

4. RADICAL POINTS OF VIEW

It should be noted, that all our troubles are due to one of the following implicit assumptions:

(i) The use of σ-additive measures in order to characterize the influence of a coalition.

(ii) The unrestricted use of the axiom of choice (AC).

As neither of these assumptions can be motivated by purely economic arguments they are open for modification in order to give a more satisfactory picture of our affairs. There are also other economic problems (see Skala (1981)) which seem to have more appealing solutions when we do not adhere to (i) and (ii).

In order to give the reader some information about the interrelationships between the axiom of choice and a few statements often used in measure theory we draw the following diagram where equivalent statements are related by a 2-way arrow. Irrevesible implications are connected by a 1-way arrow. For a thorough discussion see Pincus (1974).

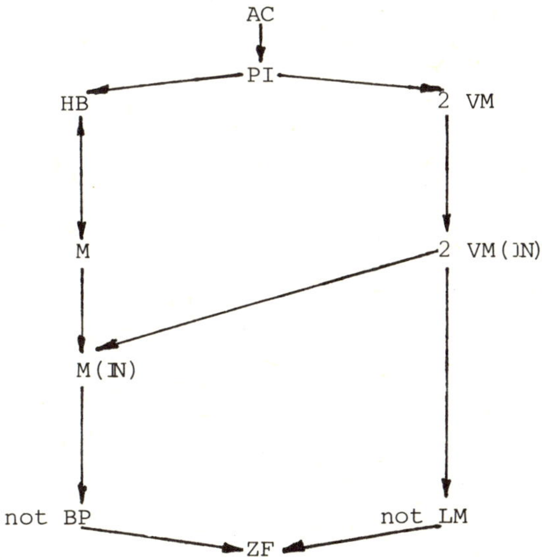

PI ...... The prime ideal theorem for Boolean algebras.
HB ...... The theorem of Hahn-Banach.
M ...... Every Boolean algebra has finitely additive probability measure.
M(ℕ) .... The power set of the natural numbers ℕ has a finitely additive probability measure which is 0 on finite sets.
2 VM .... Every Boolean algebra has a two-valued finitely additive probability measure.
2 VM(ℕ) . The power set of the natural numbers has a two-valued finitely additive probability measure which is 0 on finite sets.
LM ...... Every set of reals is Lebesgue-measurable.
BP ...... Every set of reals has the Baire property.

I. Finitely additive measures

In order to avoid the troubles which non-measurable sets cause in our model we may simply decide to be content with a finitely additive measure. If we want to recover the classical Nash equilibria, then using a Hahn-Banach type argument it is easy to extend Lebesgue integration such that every bounded function $g : [0,1] \to \mathbb{R}$ is integrable over any subset of $[0,1]$.

Moreover, denoting the extended integral by $\hat{\int}$ and the upper and lower Lebesgue integrals by $\int^*$ and $\int_*$ respectively the following holds:

$$\int_{*B} g\,dp \leq \hat{\int}_B g\,dp \leq \int^*_B g\,dp, \qquad B \subset [0,1],$$

and

$$\hat{\int}_B g\,dp + \hat{\int}_C g\,dp = \hat{\int}_D g\,dp$$

whenever $D = B \cup C$ and $B \cap C$ has Lebesgue measure 0.

It should be noted that in every Hahn-Banach argument the axiom of choice is involved in a very strong way as there is no known special case of AC which implies HB.

Although formally we are done the use of HB has, unlike Shapley's procedure, no economic interpretation. Perhaps somewhat more satisfactory in this respect is the use of non-standard analysis which has been created by Robinson (1966) and applied to large economies by Brown and Robinson (1976). (Compare also our discussion in section 2.) The principal conceptual advantage is that "large" societies can formally be considered as finite. If we want to model a society where the influence of a single agent is negligible we just have to take a "large" set which formally has the property of a finite set and attach to each member an infinitesimal weight. The integral of any function over any set then can be defined by multiplying the weight of each point times the value of the function and summing over all points in the set. In particular Bernstein and Wattenberg (1969) showed that it is possible to select a set F of the non-standard unit interval such that (1) F contains all numbers of the standard interval $[0,1]$, (2) F is formally finite and (3) F is well distributed. For any $B \subset [0,1]$ a measure $p(B)$ can be defined by dividing the number of points of F which lie in B by the total number of points in F.

Condition (3) guarantees that p is infinitely close to Lebesgue measure and behaves well with respect to translation. Moreover, the non-standard integral of any bounded function $g : [0,1] \to \mathbb{R}$ over any subset of [0,1] is defined and is infinitely close to the Lebesgue integral whenever the latter exists.

II. Doing without the full axiom of choice

A. Solovay's model

Every known proof that non-Lebesgue-measurable sets exist is highly non-constructive, i.e. it uses the axiom of choice in an essential way. Let us note that the axiom of choice leads to many counterintuitive results of which the most prominent is the Banach-Tarski paradox: "It is possible to cut the unit ball of $\mathbb{R}^3$ into a finite number of pieces which can be rearranged by using only rotations and translations so that one obtains two unit balls". However, the plausibility of AC only rests on its validity for finite sets. Thus, in spite of Gödel's consistency proof of AC, many mathmaticans feel uncomfortable with this axiom. In 1963, Cohen invented a powerful method for constructing models of set theory which enabled him to prove the independence of AC. Cohen models in which AC is false but weaker choice principles hold exist in large number. Most interesting among these is Solovay's (1970) model in which every set of reals is Lebesgue-measurable:

Theorem 6. Suppose there is a model of ZFC+I, where I denotes the statement: "There is an inaccessible cardinal number". Then there exists a model of ZF in which the following statements hold:

(1) The principle of dependent choices (=DC).
(2) Every set of reals is Lebesgue measurable (=LM).
(3) Let $\{B_x : x \in \mathbb{R}\}$ be any indexed family of non-void subsets of $\mathbb{R}$. Then there exists a Borel-measurable function $g : \mathbb{R} \to \mathbb{R}$ such that $\{x : g(x) \notin B_x\}$ has Lebesgue-measure zero.

It should be noted that DC is strictly weaker than AC and allows us to make a countable number of choices where each choice depends on the preceding ones. DC implies that for every countable family of non-void sets there exists a choice function (= CAC). As in real analysis most of the positive results can be obtained by using DC only, Solovay's model is a good candidate to accept as our universe of sets. Doing this, both the full non-cooperative ideal and the modelling of our space of agents by a non-atomic σ-additive probability space is consistent - in Solovay's model non-measurable functions simply do not exist. Thus, for our purpose we do not even need to use the nice selection result of Theorem 6(3). We would also like to mention that in Solovay's model there exists no finitely additive probability measure on the power set of $\mathbb{N}$ which vanishes on finite sets. This shows that in Solovay's model HB does not hold in its general form.

Let us summarize the situation we face.

(i) In a poor universe of sets where only such sets are allowed to exist which are needed in order to get no contradiction - i.e., Gödel's constructible universe where AC and the generalized continuum hypothesis hold automatically - there is no set such that on the power set of it there can be defined a σ-additive probability measure vanishing on singletons. (This follows from a result of Scott (see Drake (1974)) who showed that the axiom of constructibility implies that there exists no (two-valued) measurable cardinal.) Thus, either we must give up the full non-cooperative ideal or content ourselves with finitely additive measures.

(ii) At the cost of assuming the existence of an inaccessible cardinal we may consistently model our space of agents by the unit interval endowed with the Lebesgue-measure and yet preserve the non-cooperative ideal.
Using DC it is easy to prove that LM implies the existence of (two-valued) measurable cardinals (Mycielski (1964)). This shows that Solovay's model provides really a very rich universe of sets.
Without assuming I the following weaker consistency result has been proved by Solovay: If ZFC is consistent, then ZF + DC + "not every set of reals is Lebesgue-measurable but there exists a σ-additive translation-invariant extension of the Lebesgue-measure" is consistent.

When discussing Shapley's problem from the point of view of measure extensions we shall use a result of Solovay (1970) closely related to Theorem 6:

Theorem 7. Suppose that there is a model of ZFC + I. Then there is a model of ZFC in which the following statements hold:

(1) The generalized continuum hypothesis (=GCH).

(2) Every set of reals definable from a countable sequence of ordinals is Lebesgue-measurable.

(3) Let $B \subset \mathbb{R} \times \mathbb{R}$ be definable from a countable sequence of ordinals and suppose that for all x there exists a y such that $\langle x,y \rangle \in B$. Then there is a Borel-measurable function $g : \mathbb{R} \to \mathbb{R}$ which is definable from a countable sequence of ordinals such that $\{x : \langle x,g(x) \rangle \notin B\}$ has Lebesgue-measure zero.

Remark. Theorem 7 partially answers a question posed by Aumann (1969). He showed that the Lebesgue-measurability of all PCA (Σ^1_2) subsets of $\mathbb{R}$ would imply the measurable utility theorem without any connectedness assumption. As in Gödel's constructible universe there are non-Lebesgue-measurable PCA sets, we can only hope for an independence result. An application of Martin's axiom (Martin and Solovay (1970)) shows without any large cardinal assumption that ZFC + "every PCA-set is Lebesgue-measurable" is consistent. This answers question 1 of Aumann. As every projective set is definable from a countable sequence of ordinals Theorem 7 proves the consistency of ZFC + GCH + "every PCA-set is Lebesgue-measurable" if the existence of an inaccessible cardinal is assumed. Whether this result can be obtained without any large cardinal assumption is an open problem.

B. The axiom of determinateness

It is a classical theorem due to von Neumann and Zermelo that every finite two-person zero-sum game with perfect information is strictly determined, this means, that any such game has a value which can be achieved by pure strategies. Even if this statement is not so simple as the axiom of choice one may be tempted to postulate its validity also for infinite games. The motive would be the same as with the axiom of choice.

Moreover, one may argue that the result of a play between two players, each endowed with an infinite memory, should not depend on chance if complete information is assumed.

Let us associate with each subset $A \subset \mathbb{N}^{\mathbb{N}}$ ($\mathbb{N}^{\mathbb{N}}$ may be identified with the real numbers $\mathbb{R}$) the following game G_A:

The two players I and II choose successively natural numbers $a_1, b_1, a_2, b_2, \ldots$ if the resulting sequence $\langle a_1, b_1, a_2, b_2, \ldots\rangle$ belongs to A, then I wins; if it is in $\mathbb{N}^{\mathbb{N}}-A$, then II wins.

A strategy s for player I is a function which associates to every finite sequence of natural numbers a natural number. It is called a winning strategy if playing according to it, i.e. $a_n = s\langle b_1,\ldots,b_{n-1}\rangle$, results in a sequence which belongs to A, no matter what II plays. For player II a winning strategy is defined in an analogous way. We call the game G_A determined if one of the players has a winning strategy.

The axiom of determinateness (= AD) which has been introduced by Mycielski and Steinhaus as an alternative to AC simply reads: "For every subset $A \subset \mathbb{N}^{\mathbb{N}}$, the associated game G_A is determined".

By using AC Gale and Stewart (1963) proved that there are games which are not determined. Thus, AC and AD contradict each other. However, the countable axiom of choice (CAC) is easily shown to be implied by AD.

For many purposes of analysis CAC is enough. For example, the σ-additivity of the Lebesgue measure can be derived by it. We refer the reader to Mycielski (1964, 1966) where AD is studied in large detail and to Martin (1975) where the determinateness of games associated to Borel sets has been established.

With respect to our problem AD is an appealing working hypothesis as it implies the Lebesgue-measurability of every set of reals (Mycielski and Swierczkowski (1964)). Unfortunately, up to now it is not known whether ZF + AD is consistent.

5. MEASURE EXTENSIONS

In a recent paper Bierlein (1981) proposed to look at our problem from a more modest point of view. Let $(A,\mathbb{A},p)$ be a probability space and $f : A \to [0,1]$. Which conditions must be satisfied in order that there exists an extension of p which makes f measurable?

In the following we shall be concerned with conditions

depending only on A,$\mathbb{A}$ and f. Thus, the results hold for an arbitrary p.

By $\mathbb{B}$ we always denote the Borel-sets of the space in question and by $\hat{\mathbb{B}}$ the universally measurable sets. For the σ-algebra generated by a set system $\mathbb{A}$ we write $\sigma(\mathbb{A})$ and $\mathbb{K}_f$ denotes $f^{-1}(\mathbb{B}[0,1])$.

In the following we shortly sketch an alternative proof of Bierlein's result which easily generalizes if other selection theorems are available than the one essentially due to von Neumann:

<u>Theorem 8.</u> Let X and Y be Polish spaces (for example $\mathbb{R}$ with its usual topology is Polish) and let Q be an analytic subset of XxY. Then there exists a $(\sigma(\text{Anal}(X)),-\mathbb{B}(Y))$-measurable selector of Q, i.e. a function into Y which is defined on the projection of Q to X such that the inverse image of each Borel-set of Y belongs to the σ-algebra which is generated by the analytic subsets of X.

(For a slight generalization of Theorem 8 see Kurz (1977).)

From Theorem 8 we may derive a useful corollary which is essentially due to Sainte-Beuve (1974):

<u>Corollary:</u> Let X and Y be Polish spaces and let A be an analytic subset of X. To every Borel-measurable function $f: A \to Y$ there exists a function $s: f(A) \to A$ such that

(i) $f \circ s(y) = y$ for all $y \in f(A)$,

(ii) $s^{-1}(B) \in \hat{\mathbb{B}}(f(A))$ for all $B \in \mathbb{B}(A)$.

Using the above corollary it is not hard to obtain the following theorem which has been stated by Landers and Rogge (1974) for Lusin-spaces (standard measurable spaces).

<u>Theorem 9.</u> Let $(A,\mathbb{A})$ be a Suslin-space and let $\mathbb{A}_o$ be a countably generated sub-σ-algebra of $\mathbb{A}$. Then every probability measure on $\mathbb{A}_o$ can be extended to $\mathbb{A}$.

An image measure argument results in Bierlein's

<u>Theorem 10.</u> Let $(A,\mathbb{A},p)$ be a probability space and let $f : A \to [0,1]$ be a function such that

(i) $\mathbb{A} \subset \mathbb{K}_f$

(ii) $\mathbb{A}$ is countably generated

(iii) f(a) is an analytic subset of [0,1].

Then every probability measure can be extended to $\mathbb{K}_f$.

Let us note that Theorem 8 is in a certain sense best possible. Although Kondô's theorem, which we shall state presently, is available we cannot generalize Theorem 8 for arbitrary Σ_2^1-sets Q.

The reason simply is that Gödel (1938) showed that it is consistant with ZFC that there exist non-Lebesgue measurable Σ_2^1-sets.

It is therefore very interesting that Srebrny (198?) managed to generalize Theorem 8 to a wide class of Σ_2^1-sets.

<u>Definition.</u> Let X and Y be Polish spaces. $Q \subset X x Y$ (considered as relation) is absolutely total Σ_2^1 (at Σ_2^1) if there is a Σ_2^1-formula ϕ and a parameter a such that:

(i) $\forall x,y\,[Q(x,y) \leftrightarrow \phi(x,y,\alpha)]$.

(ii) For every countable transitive model M of ZFC with $\alpha \in M$, $M \models \forall x \exists y \;\; \phi(x,y,\alpha)$.

Roughly speaking, this means that every Σ_2^1-set which can be constructed and about which one can show that its x-sections are non empty within the framework of ZFC using a given parameter $\alpha \varepsilon \mathbb{N}^{\mathbb{N}}$ is at Σ_2^1.

By applying methods introduced by Solovay (1970), Srebrny obtained the following result in ZFC:

<u>Theorem 11</u>. Let X and Y be Polish spaces. If $Q \subset X x Y$ is at Σ_2^1, then it admits an $\mathbb{A}$-measurable selector for any Solovay algebra $\mathbb{A}$. (For example the Baire sets or the Lebesgue measurable sets form a Solovay algebra.)

5.1. Results Depending on Large Cardinal Assumptions

We have already used a large cardinal assumption in Theorems 6 and 7. Let us now review some other results which may be useful in our context. The most direct assumption would be to assume that the continuum is real valued measurable. Even if we assume the weaker statement that a real valued measurable (r.v.m.) cardinal $\kappa \leq 2^{\mathbb{N}}$ exists we obtain the following appealing result by an image measure argument:

Theorem 12. Assume that there exists a r.v.m. cardinal $\kappa \leq 2^{\mathbb{N}}$. Every Borel measure on a standard space A can be extended to an κ-additive measure on the power set of A.

We already mentioned Kondô's selection theorem which we could not apply without an hypothesis which assures that every Σ^1_2-set is Lebesgue measurable. In 1969 Solovay proved that the existence of a two valued measurable (t.v.m.) cardinal implies that every Σ^1_2-set of reals is Lebesgue measurable. (Even a weaker large cardinal assumption will do.) As essentially the same proof shows that every Σ^1_2-subset of a Polish space is universally measurable we can apply a deep result due to Kondô (1939) in order to generalize our Theorems.

Theorem 13. (Kondô) Let X and Y be Polish spaces and let $Q \subset X \times Y$ be a Π^1_1 (Σ^1_2)-set. In ZF it can be proved that there exists a selector of Q such that the graph of it is Π^1_1 (Σ^1_2).

Thus, assuming the existence of a t.v.m. cardinal we can generalize our measure extension theorems. Let us only state one of them:

Theorem 14. Assume that there exists a measurable cardinal. Let A be a Σ^1_2-subset of a Polish space and let $\mathbb{A}_0$ be a countably generated sub-σ-algebra of $\mathbb{B}(A)$. Then every probability measure on $\mathbb{A}_o$ can be extended to $\mathbb{B}(A)$.

We remember that the consistency result stated in Theorem 7 provides both ingredients which are needed for a generalization of the Corollary stated above, namely a selection theorem and the universal measurability of a large class of sets. As appropriate modifications show that the statements of Theorem 7 hold for Polish spaces we obtain consistency results from which we state the one corresponding to Theorem 9.

Theorem 15. If ZFC + I is consistent, then there is a model of ZFC in which the following statement holds: Let A be a subset of a Polish space which is definable from a countable sequence of ordinals and let $\mathbb{A}_o$ be a countably generated sub-σ-algebra of $\mathbb{B}(A)$. Then every probability measure on $\mathbb{A}_o$ can be extended to $\mathbb{B}(A)$.

Let us mention that Martin's Axiom which is not a large cardinal assumption also implies the measurability of Σ^1_2-sets.

Theorem 16. (Martin and Solovay (1970)). If MA and $2^{\mathbb{N}} > \aleph_1$, then every Σ^1_2-set of reals is Lebesgue-measurable.

A simple modification of Theorem 16 together with the selection theorem of Kondô can be used in order to obtain

Theorem 17. Assume MA and $2^{\mathbb{N}} > \aleph_1$. Let A be a Σ^1_2-subset of a Polish space and let space and let $\mathbb{A}_o$ be a countably generated sub-σ-algebra of $\mathbb{B}(A)$. Then every probability measure on $\mathbb{A}_o$ can be extended to $\mathbb{B}(A)$.

5.2. On Determinateness Assumptions

In a certain sense the axiom of determinateness acts like a large cardinal assumption. It should be noted that in ZF + AD the measure extension approach becomes trivial because in ZF + AD every subset of a Polish space is universally measurable. However, there is a weakening of AD by which we may obtain interesting measure extension results. Assume that we demand only games which are associated to projective subsets of $\mathbb{N}^{\mathbb{N}}$ to be determined and denote this statement by PD (=projective determinateness). Many mathematicians believe PD to be much less suspect than AD and accept it as powerful working hypothesis. For our purposes we observe that Mycielski and Swierczkowski (1964) have proved that every projective subset of the reals is Lebesgue-measurable, if one assumes PD. Essentially the same proof can be used in order to show that every projective subset of a Polish space is universally measurable. Fortunately there is also an appropriate selection theorem available.

Theorem 18. (Moschovakis (1971)). Assume PD. Let X and Y be Polish spaces and let $Q \subset X \times Y$ be a projective set. Then there exists a selector of Q such that the graph of it is a projective subset of XxY. In particular, if Q is Π^1_n (n odd) then there is a selector with a Π^1_n graph, and if Q is Σ^1_n (n even) then there is a selector with a Σ^1_n graph.

Using Theorem 18 together with the universal measurability of projective sets we obtain

Theorem 19. Assume PD. Let A be a projective subset of a Polish space and let $\mathbb{A}_0$ be a countably generated sub-σ-algebra of $\mathbb{B}(A)$. Then every probability measure on $\mathbb{A}_0$ can be extended to $\mathbb{B}(A)$.

Department of Statistics
University of Paderborn
FRG 4790 Paderborn

ACKNOWLEDGMENT

The author wishes to thank Professors Dieter Bierlein and Reinhard Selten for conversations relating the content of this paper and Professor Karel Prikry for providing him with a copy of his 1977 paper as well as with a copy of a paper of Koumoullis.

REFERENCES

Aumann, R.J.: 1964, 'Markets with a continuum of traders', Econometrica 32, 39-50.

Aumann, R.J.: 1967, 'Measurable utility and measurable choice theorem', in: La Décision, Aix-en-Provence, 15-26.

Bernstein, A.R., F. Wattenberg: 1969, 'Nonstandard measure theory', in: Applications of Model Theory to Algebra, Analysis and Probability, New York, 171-185.

Bierlein, D.: 1978, 'Measure extensions according to a given function', Lect. Notes, in: Econ. and Math. Systems 157, 25-35.

Bierlein, D.: 1981, 'A measure theoretic aspect of game theory', in: Game Theory and Mathematical Economics, Amsterdam, 399-406.

Brown, D., A. Robinson: 1974, 'Nonstandard exchange economics', Econometrica 43, 41-55.

Drake, F.R.: 1974, 'Set Theroy', Amsterdam.

Dubey, P., L.S. Shapley: 'Noncooperative General Exchange with a Continuum of Traders', Report No. 51/79 of the Hebrew University of Jerusalem.

Gale, D., F.M. Stewart: 1953, 'Infinite games with perfect information', Ann. Math. Studies 28, 245-266.

Gödel, K.: 1938, 'The consistency of the axiom of choice and the generalized continuum hypothesis', Proc. Nat. Acad. Sci. 24, 556-557.

Kondô, M.: 1939, 'Sur l'uniformization des complêmentaires analytique et les ensembles projectifs de la seconde classe', Japan J. Math. 15, 197-230

Koumoullis, G.: 1981, 'Perfect measures', Trans Am. Math. Soc. 264, 521-537.

Kunen, K.; 1980, 'Set Theory', Amsterdam.

Kurz, A.; 1977, 'Uniformisierung analytischer Mengen und eine Anwendung bei der Maßfortsetzung', Archiv Math. 29, 204-207.

Landers, D., L. Rogge: 1974, 'On the extension problem for measures', Z. Wahrscheinlichkeitstheorie verw. Gebiete 30, 167-169.

Loeb, P.A.: 1975, 'Conversion from nonstandard to standard measure spaces and applications in probability theory', Trans. Amer. Math. Soc. 211, 113-122

Łoś, J., E. Marczewski: 1949, 'Extensions of measures', Fund. Math. 36, 267-276.

Marczewski, E., R. Sikorski: 1948, 'Measures in non-separable metric spaces', Coll. Math. 1, 133-139.

Martin, D.A.: 1975, 'Borel determinacy', Ann. Math. 102, 363-371.

Martin, D.A., R.M. Solovay: 1970, 'Internal Cohen extensions', Ann. Math. Logic 2, 143-178.

Moschovakis, Y.M.: 1971, 'Uniformization in a playful universe', Bull. Am. Math. Soc. 77, 731-736.

Mycielski, J.: 1964, 'On the axiom of determinateness', Fund. Math. 53, 205-224.

Mycielski, J.: 1966, 'On the axiom of determinateness II', Fund. Math. 59, 203-212.

Mycielski, J.: 1981, 'Analysis without actual infinity', J. Symb. Logic 46, 625-633.

Mycielski, J., S. Swierczkowski: 1964, 'On the Lebesgue measurability and the axiom of determinateness', Fund. Math. 54, 67-71.

Pincus, D.: 1974, 'The strength of the Hahn-Banach theorem', in: Victoria Symposium on Nonstandard Analysis, Berlin-Heidelberg-New York, 203-248.

Prikry, K.: 1977, 'On images of the Lebesgue measure', handwritten notes.

Putnam, H.: 1980, 'Models and reality', J. Symb. Logic 45, 464-482.

Reinhardt, W.N.: 1974, 'Remarks on reflection principles, large cardinals, and elementary embeddings', in: Axiomatic Set Theory II, Providence, Rhode Island.

Robinson, A.: 1966, 'Non-Standard Analysis', Amsterdam.

Sainte-Beuve, M.F.: 1974, 'On the extension of von Neumann-Aumann's theorem', J. of Funct. Anal. 17, 112-129.

Shoenfield, J.R.: 1975, 'Martins's axiom', Am. Math. Monthley 82, 610-617.

Skala, H.J.: 1981, 'On the foundations of the social ordering problem', in: Game Theory and Mathematical Economics, Amsterdam, 249-264.

Skala, H.J.: 'Some problems of measure theory which are related to economic theory', to appear in Stochastica.

Solovay, R.M.: 1969, 'On the cardinality of Σ^1_2-sets of reals', in: Foundations of Mathematics, Berlin-Heidelberg-New York, 58-73.

Solovay, R.M.: 1970, 'A model of set theory in which every set of reals is Lebesgue measurable', Ann. Math. 92, 1-56.

Srebrny, M.: 'Measurable Selectors of PCA Multifunctions', Diss., Warszawa.

Vopěnka, P.: 1979, 'Mathematics in the Alternative Set Theory', Leipzig.

Wagner, D.H.: 1977, 'Survey of measurable selection theorems', SIAM J. Control and Opt. 15, 859-903.

A. Sochor

THE ALTERNATIVE SET THEORY AND ITS APPROACH TO CANTOR'S SET THEORY

This paper is compiled from articles of members of the Prague seminar on set theory. The first section is P.Vopěnka's lecture at the occasion of Bolzano's commemoration, Topological Symposium 1981, Prague; the second one was published in [So 1] and the third section is a part of the appendix of [V']. For the last section we selected from many papers a series of mathematical results reached in the alternative set theory and the articles in question are cited in that section.

In the first section the basic philosophical assumptions of the classical set theory are investigated (including the differences in Bolzano's and Cantor's approaches). In the second section we explain the basic ideas of the alternative set theory and to show intuitive reasons for their acceptance. Moreover we give some examples how it is possible to use methods of the alternative set theory, e.g. to explain some classical philosophical paradoxes. The third section is devoted to the connection of both these conceptions of set theory. In the last section some results of mathematics in the alternative set theory are shown and, in contradiction to previous sections, we lay emphasize on mathematical exactness (additional motivations, statements and all proofs can be found in the articles in question).

The third section should be read after the previous ones, for in the understanding of the last section the second one is useful; other parts are independent.

1. Bolzano as a source of inspiration for contemporary mathematics.

Mathematics based on Cantor's set theory is going through a crisis. The ideas on which that theory is based are inadequate to decide the truth of wide range of assertions. This crisis of truth is beginning to encroach on the mathe-

H. J. Skala, S. Termini, and E. Trillas (eds.), Aspects of Vagueness, 161–203.

mathical disciplines based on Cantor's theory.

It is possible to relegate contemporary infinite mathematics to formalism, reducing it to a search for more and more tautologies of the predicate calculus. Alternatively, we could develop many varieties of mathematics, that are all true, or all false, while affecting an enigmatic expression. But none of these approaches is strong enough to dispose of the questions raised by Cantor's theory of sets.

If we wish to answer these questions, we must look to the very roots of set theory. But this means returning to Bolzano, with whom the theory originated. Bolzano explicitly describes the points that must be considered in creating a theory of sets. The later Cantor's theory did not take many of these into account. We shall see that it was a mistake not already to have been inspired by Bolzano at the time of the first crisis, the appearance of the paradoxes.

(1) Possibility.

In mathematics, "possible" was originally understood to mean "actualizable", capable of being brought into existence. To actualize something requires a tool. There may be such obstacles that no tool will suffice. This, in classical geometry, it was asserted that four mutually perpendicular lines are not possible. That is, such lines are not actualizable, since space itself stands in the way, refusing to grant them a place.

Bolzano conceives possibility in a different and broader sense. In § 14[B] he wrote: "That which contradicts a pure, conceptual truth we must call impossible; the possible is that which is not in disagreement with any conceptual truth." Possible in this sense we would today call non-contradictory. An object is non-contradictory if the notions it involves, and the relations they lead to, do not involve a contradiction.

Thus actualizability resides in power, non-contradiction only in reason.

(2) "Object" in Bolzano's set theory.

The only objects in Bolzano's set theory are existing sets. For a set to exist, all its elements must exist, each individually. So an infinite set is infinite in the actual sense.

The first aim of Bolzano's set theory is to prove the existence of infinite sets. In the real world, it is impossible to exhibit an infinite set. In order to produce such a set, we must abandon the real world. Bolzano therefore takes recources to theological considerations. He proves the existence of a certain infinite set in the "Mind of God", and this is a manner not lacking in either charm or ingenuity.

It is of course possible to study existing sets in different ways. We can also consider sets that are merely possible, i.e. whose existence we have not proved. But then we must express ourselves accordingly.

(3) "Objects" in Cantor's set theory.

The objects in Cantor's set theory are sets that from Bolzano's point of view are merely "possible". The original intention of Cantor's theory was the study of all possible sets. Strictly speaking, the delicate nature of the objects of the theory calls for an adequate approach to their study. More precisely, Cantor's sets should be studied by means of a suitable modal logic in the Cantor theory, however, we do not study sets as possible, but as existing, as is obvious from the fact that we work directly with the predicate calculus. This must mean that we presuppose that the sets under consideration were led from possibility to existence by some completed act.

(4) Russell's paradox.

Assume that some completed act brings into existence some sets. We consider the set M whose elements are precisely the existing sets not containing themselves as elements. The set M does not exist, by the usual argument.

When we formulate this reasoning with care (and how else should it be formulated) we see that there is no contradiction whatever. All we have proved is that the set M does not exist, that is, is not among those sets which had been brought into existence by our completed act. We have not shown by this that M is contradictory. We have no reason to assume that a further act could not bring M into existence. Of course, after this second act we can proceed just as before to describe a set which does not exist. But it will be different from M.

Thus, Russell's paradox has been transformed to the assertion that in one completed act, we cannot create all possible sets.

(5) The set universe.

We shall say that sets brought into existence by a single completed act form a set universe. This, in Cantor's set theory, if we employ the classical predicate calculus, we do not study all sets, but only some set universe. Before we begin, our sets are only possible, but not existing. We begin by creating a set universe by a single completed act. By another completed act, we can obtain a more extensive universe, and view the compositum of these two completed acts as a single completed act. A set exists if all its elements exist, each individually. Thus, we can bring a set into existence only if its elements are brought into existence with it simultaneously or if they have been brought into existence by a previous completed act. So the set universe is transitive with respect to the relation "belonging to".

(6) Actualizable sets.

That which is actualizable is non-contradictory. Hence each actualizable set is non-contradictory. If a set is actualizable, then all its elements can be brought into existence by a single completed act.

Since by design of Cantor's theory is the study of the largest possible number of sets, we impose no other condition on the actualizability of a set than these two, which are necessary.

So a set is actualizable if it is non-contradictory, and all its elements can be simultaneously brouhgt into existence by a single completed act.

If two sets are actualizable, we consider their intersection, difference, cartesian product, unordered pair, etc., as non-contradictory, and therefore actualizable.

In addition, we assume that some infinite set, for example, the set of all natural numbers, is actualizable.

(7) Potential superinfinity.

Every actualizable set can be brought into existence. On the other hand, we transform the Russell paradox to the assertion that it is not possible to bring all actualizable sets into existence by a single completed act. That is, our approach to the family of all actualizable sets resembles the attitude of the mathematician who does not believe in actual infinity to the set of all natural numbers - namely, that the set of all natural numbers cannot be created, but each individual natural number can be, if we omit the technical difficulties involved.

The family of all actualizable sets is larger than any infinity it is possible to create. Therefore we will call it superinfinity. This superinfinity is not actual, and it is not possible to actualize it by a single completed act; therefore, we consider it only in the potential sense.

The theory of sets, whose aim was the actualization of potential infinity is not able to abolish the potential infinity, but merely shifts it to a higher level.

(8) The assumption that the power set is actualizable.

Let X be an infinite existing set. By the power set of X,P(X), we mean the set whose elements are precisely the actualizable subsets of X. By the condition imposed on actualizability, each non-contradictory subset is actualizable. Thus, P(X) is the set whose elements are precisely all non-contradictory subsets of X.

At present, we have no reason to think that P(X) is contradictory. However, the problem of actualizability is more complex. To view P(X) as actualizable, we must hold that all the actualizable subsets of X can be brought into existence by a single completed action.

Cantor's set theory admits P(X) as actualizable. This places P(X) outside the framework of Bolzano's set theory. At the same time, Cantor crosses what was at that time uncrossable border of mathematics, and leaves geometric space. His sets will not fit in space, and the role which up to then had been played by space was now to be played by Cantor's set theory. This is all a consequence of Cantor's famous theorem that the sets X and P(X) have different cardinals.

We will say that a set universe is perfect if for each X lying in it, each actualizable subset of X also lies in it.

(9) The basic hypothesis of Cantor's set theory.

We now know that if we study Cantor's set theory by classical methods, we do not study all possible sets, but only the sets from some set universe. That does not mean that we want to give up the study of all actualizable sets. We assume that the properties of the family of all actualizable sets are reflected in suitable set universes. In particular, we hold that so-called ZF-universes are suitable. We say that a universe of sets is a ZF-universe if its sets satisfy all the Zermelo-Fraenkel axioms.

The basic hypotheses of Cantor's set theory amounts to this: that it is essentially possible to reflect the family of all actualizable sets in a suitable ZF-universe. Then we can more accurately express the subsequent procedures.

If X is an actualizable set, then it is possible to bring into existence a ZF-universe containing X by a single completed act.

(10) Concerning a formulation of Bolzano.

In [B] Bolzano writes: "Let us now proceed to the consideration of the most remarkable peculiarity that can occur in the relation of two sets if both are infinite, that even in fact must always occur, but has been up to now unfortunately overlooked..."

"Namely, I assert: any two sets, both infinite, could be so related that it is possible to pair each element of one with an element of the other in such a way that no element in either set is unassociated, and none appears in two or more pairs".

Today we would say more concisely that any two infinite sets can be put into a one-one correspondence. Knowing that

Bolzano distinguishes between existing sets and possible sets, we are not surprised that he says the sets "could" be so correspondent, not that they are. Such a correspondence is only non-contradictory; whether it is also realizable is another question.

Under the influence of later results of Cantor's set theory, mathematicians viewed the assertion that two infinite sets can always be so related as Bolzano's error.

(11) General collapse.

By general collapse we mean the following assertion: If X and Y are two existing infinite sets, then a one-one correspondence between X and Y is actualizable.

Today we can view the consistency of general collapse as sufficiently established. We certainly agree that the set of bijections whose domain is a finite subset of X and whose image is a finite subset of Y is non-contradictory. The natural order on this set forces (in Cohen's sense) the existence of one-one mapping from X onto Y.

This argument shows that F is non-contradictory, i.e., if we consider such a correspondence, we will not be led to a contradiction. Since the cartesian product of X and Y is assumed to be actualizable, there is no alternative but to consider the correspondent F to be actualizable.

(12) Consequences of general collapse.

If X is an existing infinite set, then the set P(X) is not actualizable. If we bring into existence an infinite set Y whose elements are subsets of X, then by general collapse we can actualize a one-one correspondence F between X and Y. Then the Cantor diagonal method tells us how to find a subset of X that is not an element of Y. The set P(X) is not actualizable, since it is not possible by a single completed act to bring into existence all the actualizable subsets of X at once. At the same time, of course, the set P(X) may be non-contradictory, so we can study it - not, however, with the classical predicate calculus, but by means adequate to its character.

From the above considerations it follows immediately that

a ZF-univese containing an infinite set cannot be perfect.

At the present time, the set of real numbers cannot be considered as completed. By external means, i.e. using procedures outside the framework of the set universe we are working in, we can number all the existing reals by integers. Then, again by external means, we can produce a new real number.

Similarly, all the points lying on a given line, that is, all existing points, do not exhaust the line. We can always add to these points another point. So it is not true that the line is the set of all points lying in it. However, it is determined by that set.

(13) Cantor's set theory.

Mathematics whose world is Cantor's set theory is mathematics that takes place in some ZF-universe. That universe is not perfect; of course we do not notice this when we work inside it, that is, when we view as actualisable only what is actualized in it.

The construction of Boolean models, permutation models, Gödel's L-model and the like which we carry out inside our universe suggest that there are other possible universes. Indeed, if we look at these constructions in terms of general theory of actual infinite sets introduced above, we see that they provide a method for actualizing other ZF-universes, broader or narrower than our own. Since the power set of an infinite set is not actualizable, it is not possible in any absolute sense to pose the question of the truth of the axiom of choice, the continuum hypothesis, and the like. The only thing that makes sense is to assert their truth in this or that ZF-universe. However, that is a topic in itself, and clearly one where many satisfactory answers are known.

The mathematics of Cantor's set theory roots in the basic hypothesis of that theory; it is now clear that we must abandon the assumption that the corresponding universe is perfect. But the arguments that earlier gave evidence of the benefits of that hypothesis now lose their weight. In particular, the assumption of the power set axiom is not justified, even if we understand by it only the relative power set. Mathematics based on Cantor's set theory is losing the ground under its feet.

With the assumption that the power set of an infinite existing set was actualizable, mathematics entered a peculiar world. It is a world only in the sense of a fairy tale, in which we meet a large number of wondrous creatures in the form of fantastic topological spaces, algebraic structures and the like. The immortality of Cantor's theory of sets lies not in its truth but in just the opposite - in its fantasticness and its suggestivity, by which it came to captivate the mathematics of an entire epoch.

(14) The problem of the compatibility of the familiy of actualizable sets.

As soon as we sever the non-contradictory sets from the actualizable, we must deal very carefully with the family of actualizable sets. If we are thinking about two non-contradictory sets X and Y, then reason does not forbid us to think about both at once. However, a completely different situation presents itself if X and Y are actualizable sets. We have no guarantee at all that the family of actualizable sets is compatible, i.e., that if each of the sets X and Y is actualizable, it follows also that they are actualizable together. In the real world something similar occurs. Consider, for example that it is possible to carve several different statues from one stone, however - as soon as one is carved it is impossible to carve many of the others. There is much that indicates that the family of actualizable sets is not compatible. In any case, until we have settled this question, we must deal with it as though it were not compatible.

When we bring a set into existence, we begin to "filter" the family of actualizable sets. That is, as soon as we actualize a set, we may shrink the family of realizable sets. At the next step we can actualize only sets from this smaller family, and as we continue the process, the family gets even smaller. So we may say that the process of actualizing passes the actualizable sets through a filter.

(15) The set of natural numbers.

We hold the natural numbers in the mind through notions and relations between them that we created for this purpose, and which we belive do not lead to a contradiction. At the same time, of course, we are not able to devise a conception that can grasp them with unambiguous precision. So if we assume that the set of natural numbers is actualizable, then we are assuming that a set is actualizable whose properties correspond to our notions. But this says that possibly there are other sets that have the properties ascribed to the set of natural numbers, even containing the individual numbers in von Neumann's sense.

However, we place other requirements on the natural numbers related to the family of actualizable sets. In particular we assume that this actualized set is standard, that is, that each of its proper actualizable subsets necessarily has a first element. If the family of actualizable sets is compatible, there is at most one standard set of natural numbers. However, without this assumption we cannot rule out the possibility that there may be more standard sets of natural numbers, more precisely, various actualizable sets of natural numbers, differing even with respect to the validity of an assertion about the natural numbers, may be actualizable as standard. However in such a case, if we actualize one, we cannot actualize any of the others.

Infinite sets are usually brought into existence by first actualizing a standard set of natural numbers. This imposes a filter on the family of realizable sets. This step is decisive, since through it we specify what it means to be non-contradictory.

Thus the non-compatibility of the family of actualizable sets is a very natural explanation of the undecidable assertions in the theory of natural numbers.

2. The alternative set theory

The classical set theory arose during the last century (B.Bolzano and G.Cantor). In spite of many objections at the beginning, this theory later gained general recognition and - moreover, it influenced the whole mathematics in a decisive way. This influence is not restricted to the formal aspects, in fact the classical set theory changed the whole conception of mathematics and made the rise of new mathematical disciplines possible. We can ask whether there exist possibilities to build up another theory that could replace

Cantor's set theory and, consequently, if mathematics in our century could develop in another way. This question is in no way self-purposed, it is meaningful to look for an alternative to the classical set theory at least for the following two reasons:

(a) It is not evident that Cantor's set theory is the best possible description of our comprehension of the real world and it is quite possible that another theory would enable us to grasp even real world phenomena which Cantor's set theory cannot describe in a natural way.

(b) It is by no means obvious that Cantor's set theory is the best possible framework for mathematics - e.g. if we want to develop Leibniz's idea of infinitely small quantities (infinitesimals) in the classical set theory we have to use Robinson's nonstandard methods which, however ingenious, are rather complicated and require deep mathematical maturity.

The alternative set theory, which was created by P.Vopěnka (cf.[V]), is an attempt to construct a theory that could serve as an alternative to Cantor's set theory. At first we shall try to argue that the basic ideas of the alternative set theory can be motivated by our intuition of the real world. In the second part of this section we are going to demonstrate the way through which the means of the alternative set theory can be used dealing, for example, with some classical paradoxes and with the conception of the Calculus. It is necessary to emphasize that in the alternative set theory even means which are not so naturally available in Cantor's set theory are available,e.g. infinitely small quantities. One of the specific features of the alternative set theory is that it makes possible to combine two different views of a given problem, which can be hardly accomplished so naturally in the classical set theory. The discussion of connection between discreteness and continuity may serve as an example of this attitude.

(1) First thesis - finiteness.

In real world we deal with collections describable by (writable) list of their members, e.g. the collection of all people in a given room. This type of collections will be formalized in the alternative set theory by the notion of a set. Every (writable) list contains only finitely many items and hence these considerations lead us to the first thesis of the alternative set theory:

I. There is no infinite set.

We can express this thesis more mathematically by the statement that sets in the alternative set theory fulfil the law of mathematical induction. This means that every property can be expressed by reference to sets only (!) if the empty set satisfies this property, and if the assumption that a set satisfies this property implies this property for all sets arising from the given set by adding one element, then all sets satisfy the property in question. This thesis can be reformulated in the form of the assertion that every set is formally finite. Here we used the adverb "formally" because any attempt to formalize our everyday intuition about finiteness mathematically does not yield, after all, the true picture of the intuition. In particular, we cannot exclude the possibility that even some objects which are not intuitively finite fulfil the conditions required by our formalization. Thus if we formally describe our view of the real world we cannot postulate more than this formal finiteness. Later we shall see that this pass can be used for the grasp of real situations when the collection in question is finite from one point of view but from another standpoint is very large.

We must emphasize that by accepting this first thesis we disagree with Bolzano's principle and thus we return to older conceptions of mathematics (cf.1(2)).

(2) Second thesis - existence of semisets.

Our idealizations and generalizations, e.g. the collection of all natural numbers, are not included in the type of objects described above. We never suppose that we are able to write a list of all natural numbers and therefore it would be erroneous to formalize as sets all objects which our idealizations and generalizations give to us. Hence we introduce in the alternative set theory a new type of objects, called classes, which can be imagined as formal counterparts of our idealizations and generalizations. From a rather different viewpoint we can say that classes represent our approaches to the real world.

A class is mostly understood as being described by a property - this means that in the moment of its creation it is not necessary to know all its elements (contrary to 1.(5)). For example, the property "to be a natural number" and even the class of all natural numbers can be considered before all natural numbers are created.

The addition of classes as new objects is nothing new and revolutionary since even some axiomatizations of Cantor's set theory deal with this notion. In any case the classical set theory assumes that classes which are not sets themselves, called proper classes, are enormously large and thus they cannot be subclasses of sets.

In the alternative set theory we change this point of view. Let us first motivate this opposite standpoint. Suppose that we have a large library. Of course, it is natural to assume that we are able to make a list of all books in our library and hence the collection of all books in our library will be treated as a set. However, can we make a list of all interesting books in the library? Some books are definitely interesting and others completely dull, but usually we have also books about which it is difficult to decide whether they are really interesting or not. Thus it would be very difficult to write a list of all interesting books in our library though such a list would not be so extensive as the list of all books in the library. We assume that, in principle, we are able about each book to decide whether it is interesting or not, but such a decision might be very time consuming. For example let us suppose that there is a lack of space in our library and we want to discard uninteresting books. Then we are forced to decide explicitly whether any particular book is interesting and then keep it in our library or whether it is not interesting and then to discard it. In the real situation we shall discard books as long as we have enough place; for lack of time we would approximate the collection of all uninteresting books its subcollection which can be better described. Hence we shall never be sure that just those books which are interesting remain in the library. On the other hand let us realize that it is very convenient to speak about the collection of all interesting books in our library - and we often do it. Our idealization holds this collection as already created although its extension is not precisely described.

Thus, from the above described viewpoint it seems to be suitable to accept, in the alternative set theory, the following thesis:

II. There are sets with subclasses which are not sets.

Consequently, defining (proper) semisets as those (proper) classes which are subclasses of sets, our second postulate guarantees the existence of proper semisets. We can imagine proper semisets as formalizations of collections (extensions of) which are not precisely described (which are not constituted by lists of their members). Let us realize that a great deal, almost all, of intuitive properties do not precisely describe the corresponding collections. Let us mention some other examples of such properties: "to be a good man", "to be living", and so on. In every case we can find objects about which we shall dispute whether they have the property in question or not.

Let us note that the decision whether a collection of objects of real world will be formalized as a set or as a proper semiset depends on our standpoint. For example, considering the property "to be an interesting book" we have represented the collection of all books in our library by a set. However, even the property "to be a book" is not quite precisely defined - there are manuscripts, paperbacks, reports,partly destroyed books and so on. Thus we can shift our point of view by dealing with still higher precision in description and consider the collection of all books to be a proper semiset.

We have stated that proper classes in the alternative set theory represent collections we cannot have in sight as the whole. Such collections can be divided into two types. The first type consists of collections which are very large (there are no restrictions for construction of new and new elements of such a collection) and in the second type there are collections which one is not able to survey since they are not described precisely enough (proper semisets). Let us recall that the classical set theory admits only classes which are formal counterparts of collections of the first type. We have postulated that all sets are formally finite. But sets having proper subsemisets can be imagined as "inaccessible". There are many nice examples of the fact that the conception described above approximates to our understanding of the real world; let us mention only the following one. Physics hypothesize that the number of all atoms of our Earth is finite. But this finite natural number is only formally finite - we shall never be able to recognize (let alone count) all the atoms of the Earth. The collection of recognizable atoms is a subcollection (proper semiset) of the collection of all atoms of our Earth.

A great part of mathematics deals with our idealizations and generalizations. Therefore if we want to build mathematics in the alternative set theory we really need to be able to grasp these idealizations in our theory. In fact, the alternative set theory without classes (i.e. only with the notion of a set) would be very poor and it could not serve as a framework of a conveniently rich mathematics; such a theory would coincide with the fragment of Cantor's set theory describing only hereditarily finite sets.

Let us mention that there are more branches of mathematics endeavouring to grasp real situations of the described type; let us mention, for example, fuzzy sets, Parikh's feasibility, theory of semisets and intensional logic.

(3) Third thesis - prolongation axiom.

The simplest phenomenon of infinity that we meet in the real world is unlimited processes, i.e. processes in which for every finite collection of objects of the investigated type we are able to construct a further object of the same type differing from all previous ones. We acquire experience with unlimited processes when we investigate phenomena converging toward the horizon of our observational ability. The phenomenon of infinity described above is mathematically formalized in the alternative set theory by the notion of countable class. Mathematically we define a class to be countable iff it is a proper class which can be ordered in such a way that the class of all predecessors of each its element contains no proper class. In particular the class FN of all "accessible" natural numbers (cf.4.(1)) is countable.

Imagine that we find ourselves on a long straight road lined with large stone set at regular distances and let the guard stones reach the horizon, that is we cannot distinguish any last stone. It is quite natural to assume that even beyond the horizon the road is lined by quard stones at least for some distance. Of course, this is a hypothesis which cannot be verified unless we change our position or improve substantially our observational ability. Since after any visible guard stone there is a further one, we encounter a phenomenon of infinity. We represent formally the collection (sequence) of guard stones before the horizon by a countable class. From the chosen point of view we have to formalize the collection of all guard stones by a set (which is a superset of the original countable class).

Similarly even the collection of all events of fall about we shall know that they conform to Galileo's law, has no limitations in prolongation (considering the horizon in a very general sense). Hence even this collection is treated as countable class. Moreover we hypothesize, transcending the horizon, that all falls conform to Galileo's law. We have here approximated a given countable collection by a larger collection consisting of objects with the same property in question in such a way that the formal counterpart of this larger collection was a set.

A formalization of this way of thinking can be based on an assumption that every countable class can be approximated by its superset with similar properties. Moreover it is sufficient to restrict ourselves to one property, namely to the property "to be a function" which is universal in the sense that statements talking about all properties can be proved from the statement dealing with this property only. Let us mention that we accept the usual conception of set theory in which functions are classes with the particular property. Therefore the previous considerations lead us to the following thesis:

III. Every countable function is a subclass of a function which is a set.

This thesis can be held for an expression of the human aspiration of going beyond horizons - to transcend them mentally. It appears to be the most important and powerful thesis of the alternative set theory. General mathematics works with infinity, therefore admitting only formally finite sets, but we must be able to reach infinity through finite sets. Our third thesis expresses exactly a formalization of this approach (which is, in fact, the only possible one in our case). Moreover, it makes possible to pass from investigation of infinite structures to investigation of formally finite sets with convenient structure. The classical set theory puts infinity only "beyond" finite sets while the alternative set theory puts it also "among" formally finite sets, since there are sets containing infinite subclasses.

(4) Fourth thesis - two infinite cardinalities.

In the real world we meet yet another type of infinity than countability, namely the continuum. We meet with this type of infinity whilst observing a glass of water or saying mathematically what we consider as the real line. In fact, we can formalize the collection of all atoms in the glass of water as a set. Treating this set we cannot discriminate all its individual atoms, because this lies beyond the horizon of our observational ability (the collection of all atoms that we are able to distinguish separately is represented by a countable subsemiset of the investigated set). In the alternative set theory the continuum is realized by any set with a proper subsemiset. Similarly as in the classical case, we can show in the alternative set theory that the continnum is another (larger) type of infinity than countability.

Cantor's set theory examines plenty of types of infinity. Their existence follows from the assumption that for every set the collection of all its parts constitutes a set (powerset axiom) and from the existence of at least one infinite set. On the other hand the types of infinity obtained in this way are difficult to imagine intuitively. This led some mathematicians (e.g. Poincaré) to refuse such types of infinity. Since we want to approximate our intuition in the most adequate way we restrict ourselves to countability and continuum and we do not speculate about other types of infinity. Thus we accept the last thesis of the alternative set theory:

IV. There are exactly two distinct types of infinity.

The axiomatic system of the alternative set theory consists of the formalizations of the above stated thesis and of the usual set theoretical axioms (cf.4.(1)).

Up to the end of this section we are going to show the approach of the alternative set theory to some particular classical questions.

(5) Bald man paradox. (Eubulides of Miletus)

This well-known paradox rests on the fact that there is no minimum number of hairs on the head which dictates that a man is not bald. In the conception of the alternative set theory the collection of numbers of hairs which men have who are not bald is represented by a semiset. We do not assume in the alternative set theory that every non-empty semiset of natural numbers contains the minimal element (such an assumption is accepted only for sets) and therefore this paradox is automatically resolved in the alternative set theory.

In the alternative set theory we have "inaccessible" natural numbers (let us recall that a natural number is inaccessible if there is a proper semiset consisting only of numbers smaller than it). This enables us to formalize mathematically, e.g., the following situation: Suppose we walk such a long way that it is impossible to have it uninterrupted. It is natural to assume that at every moment we are able to make one step more. Hence, formalizing the collection of all steps of uninterrupted walking, we get a countable class which is a proper subsemiset of the set which is a formal counterpart of the collection of all steps which are necessary for reaching the end of our journey.

(6) Mathematical induction.

There are many properties of finite sets which are accepted as self-evident. However, the critical analysis shows a lack of arguments for their absolute acceptance. As an example we can take the assumption that each subsequent counting of a given finite set gives us the same number of its elements. The reason for the acceptance of this assumption can be derived from our education and from the fact that we can verify the validity of this hypothesis in any case when the number of elements of the given set is relatively small. From the mathematical point of view we can prove the previously mentioned assumption by mathematical induction, but this only reduces our question to the analogous problem of validity of mathematical induction. In the alternative set theory mathematical induction in the general form holds only for "accessible" natural numbers, more specifically for these natural numbers every counting yields the same number. For "inaccessible" natural numbers every counting formalized as a set yields the same natural number but there exist proper classes which are one-one mappings of the "inaccessible" natural number in question onto its successor.

Let us note that the role of ω of the classical set theory is played in the alternative set theory by the class of all "accessible" natural numbers FN. In particular this class is countable while the class of all natural numbers has in the alternative set theory the cardinality continuum, since even every "inaccessible" natural number has this cardinality.

(7) Connection between the discrete and the continuous.

Describing such a connection is one of important problems and the basis of this matter can be found in Greek philosophy where the non-existence of motion was proved exactly by a combination of discrete and continuous views (Zeno of Elea). As far as modern physics is concerned let us mention similar problems concerning the nature of light.

The alternative set theory makes it possible to transfer the investigation of a real line, which is continuous, to the investigation of a discrete set with a convenient structure. Let us take an example: when we observe a table we can imagine it as very large but formally finite set of atoms. Each of our observations discovers some points as different. We can consider a sequence of such observations, e.g. from a distance, by the eyes, by microscope and so on, which gradually enables us to discriminate better and better points; but in any case there remain atoms which are so close to each other that we are not able to separate them (unless we consider the mental experiment of shrinking ourselves to the size of atom). We introduce the indiscernibility equivalence, which is formally a proper class, then when using it the originally discrete structure begins to appear to us continuous. Therefore we can see the table in two ways: continuous and discrete (cf.4.(6)). We define that motion in the alternative set theory is an "inaccessible" long series of steps, in each step the position of the investigated object can be changed only infinitely little (w.r.t. the indiscernibility equivalence). Moreover, let us realize that this conception of motion requires unconditionally both continuous and discrete views. There are non-trivial mathematical statements about the connection between these two approaches (cf.4.(7)).

(8) Achilles and tortoise. (Zeno of Elea)

We can again imagine the whole course as an "inaccessible" large set of atoms. This set is divided into two sets: a set of all atoms where the tortoise is ahead of Achilles and its complement. The latter set has, of course, the first element. Zeno's argument shows how to put our horizon into the set of all atoms when the tortoise is ahead of Achilles. However, when we assert that Achilles will never catch up the tortoise we deliberately make the assumption that another situation of the horizon is impossible.

(9) The alternative set theory as an alternative to Cantor's set theory.

Before Cantor's time there were several mathematical disciplines, e.g. geometry, arithmetic, the Calculus, basis of algebra and so on. These disciplines had a great number of common points and mathematicians anticipated that there must be a universal basis for all of them, such a basis was realized by set theory. (More precisely, it was shown that it is possible to model all mathematical disciplines of that time in set theory.) In other words, we can say that the classical set theory built the universal language for the whole mathematics. Set theory had still greater influence on some branches of mathematics, e.g. on the theory of real numbers.

We hope that the alternative set theory can substitute Cantor's theory in this respect. Many disciplines in question e.g. arithmetic, real numbers, the Calculus and so on, were already reconstructed in the alternative set theory.

Classical set theory made possible to grasp infinity in mathematics and investigate it. Moreover, the different aspects of infinity obtained in set theory a universal basis. (At the same time the conception of potential infinity was replaced by the conception of actual infinity.)

The last thesis of the alternative set theory made the theory of infinity in the alternative set theory rather simple; on the other hand the above mentioned aspect of the grasping of infinity was preserved in this theory. Moreover, it is a rather surprising fact that a great deal of mathematics can be built on the basis of a theory with such a simple conception of infinity.

Furthermore, let us notice that, similar to Cantor's set theory, even the alternative set theory enables development of new mathematical disciplines. Ideas of the alternative set theory make possible to constitute a new approach to the whole mathematics. We are once more making an effort to formalize mathematically our intuitive conceptions and, moreover, we want to obtain in the alternative set theory a mathematical formalization of conceptions which up to now were not adequately formalized. It is natural that the classical mathematics for a great deal passes from Cantor's set theory into the alternative set theory without any changes, but in any case we are obliged to ask, at first, whether the mathematical notion in question agrees with our intuitive conception; the mechanical application of even ingenious ideas of the classical set theory led often astray. Thus in the alternative set theory there are disciplines which correspond to the analogous ones in Cantor's set theory, but there are even disciplines which are based on specific means to the alternative set theory. Furthermore, it is necessary to emphasize that we consciously exclude from the alternative set theory such branches of mathematics which were evoked by the classical set theory itself (and which are from the point of view of the alternative set theory self-purposed).

In the alternative set theory we are able, assuming additional axioms, to imitate Cantor's set theory (cf.3.). Therefore, it is possible to transfer all results which were obtained in the classical set theory (and even all problems which were formulated there into the alternative set theory. But from the point of the alternative set theory, Cantor's set theory is by no means the only possible theory of infinite sets.

(10) The Calculus.

The fundamental notion of Leibniz's conception of the Calculus was the notion of infinitely small quantities (infinitesimals). It is well-known that this notion was, on one hand, very fruitful, but that, on the other hand, it seemed to lead to inconsistency. Therefore many famous mathematicians, e.g. d'Alambert, Bolzano, Cauchy and Weierstrass, tried to build the Calculus on a better basis - this means on an undoubtedly consistent basis. They built the modern conception of the Calculus in which the notion of infinitely small quantities is not used. However, in technical practice infinitely small quantities are used up to now because they facilitate calculations.

In 1960 A.Robinson constructed a nonstandard analysis and thus gave us a suitable framework for the development of the idea of infinitely small quantities in Cantor's set theory. Moreover he showed consistency of the Calculus based on this notion (relatively w.r.t. the classical set theory). Some remarks as to the connection between the alternative set theory and nonstandard methods can be found in [So 3]. Using similar methods - but very naturally - we can constitute the notion of infinitely small quantities in the alternative set theory (cf.[S-V 1]) and return to ideas of Leibniz and Newton without apprehension of mistakes. This approach makes even some theorems of the Calculus easier to prove. Moreover, the creation of the Calculus in the alternative set theory put before us a very natural question whether all properties expressible in Leibniz's Calculus are expressible even in the contemporary conception of the Calculus.

Let us interpret Leibniz's conception of the Calculus using the terms of the alternative set theory. Leibniz deals with the class of all real numbers R and extends it by adding new elements, so that he is able to model infinitely small quantities in this extended class Ex(R). Furthermore, he supposes, that R and Ex(R) have "same" properties. At the same time he assumes that every real function defined on R can be uniquely extended to a function defined on Ex(R). The class FN of all "accessible" natural numbers is held as a subclass of R. Using properties depending only on classes R and Ex(R) we are able to describe all notions of Leibniz's Calculus. However, it is possible to construct a property of real functions depending on classes R and Ex(R) only which cannot be expressed in the modern Calculus - even if we allow notions of functional analysis (cf.[So 2]). Fortunately we are able to describe a smaller collection of properties which contain all notions of Leibniz's Calculus - properties depending only on classes Ex(R) and FN. These properties are expressible in the current Calculus (see [Č] and 4.(15)). Moreover, P.Vopěnka constructed a simple algorithm translating some properties of the previously described type into the Calculus without infinitely small quantities (cf. [So 2]). This algorithm agrees with the translations of notions of Leibniz's Calculus into modern versions of the Calculus which were in fact used in the development of the current Calculus.

(11) Concrete natural numbers as "inacessible" ones.

The careful reader had certainly realized that the second basic thesis of the alternative set theory guarantees the existence of sets containing proper subsemisets, but we never accepted the idea that some directly described set contains a proper subsemiset. Though exactly such kind of assumptions would be a better description of our intuition. For example we would be able to estimate (by 10^{10}, say) the number of hairs which a man can have and hence, following our formalization of the bald man paradox, it would be natural to accept the axiom "there is a proper semiset contained in a set with 10^{10} elements" i.e. "10^{10} is inaccessible". However, there are very important reasons preventing us to accept such kind of assumptions; let us, at first, deal with philosophical one: The decision whether 10^{10} is "inaccessible" depends on the chosen point of view. For example, if we count adding successively one element after another, then the number 10^{10} seems to be "inaccessible" but when we count using powers then we reach the number 10^{10} very quickly. Since the alternative set theory is meant as a general theory, it cannot depend on a random choice of standpoint.

From the metamathematical standpoint we have followed our intuitive conception of the real world as far as possible whilst still preserving the consistency of our theory (in the classical sense). In fact, it can be shown that the alternative set theory based on thesis I - IV is consistent (more precisely, the relative consistency of the alternative set theory w.r.t. some fragments of the classical set theory can be demonstrated; cf.4.(19)). But if we admit a postulate of the form "10^{10} is inaccessible" the alternative set theory would lose its consistency; and arbitrary nonsense would be provable in such a theory. On the other hand even the alternative set theory with an assumption of the form "10^{10} is inaccessible" may be interesting, although inconsistent. Any mathematical proof of inconsistency is very long (it has to have about 10^{10} steps) and thence such a proof is not very conclusive. Since the alternative set theory with the investigated axiom describes certain intuition of the real world it would be correct to keep constantly this point of view and therefore even the mathematical proof in question has to be held from this viewpoint as "inaccessible".

3. Actually infinite sets from the point of view of the alternative set theory.

In this section we shall investigate Cantor's set theory from the standpoint of the alternative set theory. In 4.(19) and in [P-S] the reader can find some results concerning the alternative set theory from the point of view of model theory in Cantor's set theory.

(1) Theological motivations

Up to now we have dealt with actually infinite sets so as we would know what we are speaking about. However, actual infinity, or more precisely what is meant by actual infinity, is not a phenomenon; infinite sets are not something we would be able to perceive. It was Bolzano who knew rather well that only God can observe infinite sets. Mathematicians could not reconcile themselves with Bolzano's demonstration of the existence of an infinite set and, indeed, since they did not find other, they tried to forget it. However, the theological motivations could not be removed from mathematics entirely, but only from the consciousness of living mathematicians. They only remain more strongly in their subconsciousness. Claiming the right of deciding what is true about infinite sets, the mathematicians took a standpoint such that they could hold the view from which these sets are visible, that is the God-like position. When we say that some real number with particular properties exists we do not mean that we or anybody else is able to find it, but that God sees it. Moreover, we began to carry out constructions into which infinitely many objects enter. For example we create the limit to every sequence fulfilling the Bolzano-Cauchy condition. We often do not know anything more about the limit of such a sequence except that it exists. Infinitary constructions and manipulations are often beyond the possibility of our insight. Owing to that objects began to appear in mathematics which nobody can directly observe. Bolzano's continuous function without a derivative is the first example. Thus we create objects which, in fact, we are not able to create.

The only true motivation for acceptance of the axiom of choice is that it is certainly within the "Power of God" to create a suitable choice function. On the other hand, the axiom of determinateness is equally well motivated since it is inconceivable that God would not know how to play any corresponding game in the optimum way.

Finally, let us realize that also the acts by means of which we have brought various sets from possibility into existence in the first section were nothing else than acts of the "Power of God".

(2) The phenomenon of actual infinity.

In whatever way we would try, we are never able to see actually infinite sets through the "Eyes of God". If we look at such a set we see just that we can see and not what we would like to see. When we attempt to see an actually infinite set, i.e. both the whole set and all its individual elements, then we imagine this whole as well as the individual elements as they are spaced in various directions from us to the horizon and which fade away as they approach it.

Thus if we try to visualize an actually infinite set then we imagine only a countable class in the sense of the alternative set theory. In other words what we believe to be an actually infinite set is, in fact, just a countable class. Apparent phenomenon of actual infinity is therefore the phenomenon of a countable class.

(3) Models of filters of actualizable sets.

If we imagine the standard set of natural numbers then we imagine the class FN. In the alternative set theory, there is a class A which is a convenient model of a maximal filter of actualizable sets (cf.4.(5)). Thus if we believe that we work with actually existing infinite sets then, from the standpoint of the alternative set theory, we are inside the class A.

The extended universe (i.e. the system of sets and classes) studied by the alternative set theory depends essentially on the horizon of our observing ability. The aim of the class FN is exactly to capture the notion of distance of the horizon. Since the distance of the horizon may vary, various extended universes are possible, differing, in particular, in their classes FN. The various extended universes may, of course, differ even in some statements concerning classes. Indeed, each such extended universe has its own class A which can have, however, different additional properties in various extended universes and consequently model another filter of actualizable actually infinite sets.

(4) The axiom of Cantor's set theory.

In Cantor's set theory we assume that there is possible such an extended universe in which a class A fulfils the basic hypothesis of Cantor's set theory. Now, by a ZF-universe we mean such an element of the class A that all axioms of ZF hold on the class $u \cap A$, and furthermore, for each $x, y \in A$ the assumption $y \in x \in u$ implies $y \in u$.

The basic hypothesis of Cantor's set theory can now be fortunated as the following additional axiom of the alternative set theory:

To every $x \in A$ there exists a ZF-universe $u \in A$ so that $x \in u$.

Obviously this axiom can be strengthened in various ways in agreement with the classical Cantor's ideas. By the acceptance of this axiom, which Cantor's set theory accepts subconsciously, Cantor's set theory becomes a study of one particular class of a certain type of extended universe studied by the alternative set theory. Motivations based on the basic intuition of the alternative set theory which would give evidence for acceptance of the above axiom were not found up to now. It would be extremely interesting, indeed, if such motivations were found.

4. Mathematics in the alternative set theory.

In this section we selected some results which were obtained in the alternative set theory. The reader is referred to the papers cited for more information.

(1) Axioms of the alternative set theory and natural numbers ([V], cf. also [S 1]).

The language of the alternative set theory is the usual language of set theory, i.e. $\in$ and $=$ and one sort of variables (called classes) X,... . The unary predicate "to be a set" we define - as in GB - by $\mathrm{Set}(X) \equiv (\exists Y) X \in Y$ and we use variables x,... for sets. Semisets are defined as subclasses of sets. Set-formulas are formulas in which only set variables and constants occur; in normal formulas only set variables are quantified. We use the usual set theoretical notions.

AST (the basic alternative set theory) is the theory with the following axioms:

A1 (extensionality for classes) $(\forall X,Y)((\forall Z)(Z\varepsilon X\equiv Z\varepsilon Y)\equiv X=Y)$

A2 (existence of classes; "Morse") For <u>every</u> formula Φ of the language of set theory we accept the axiom $(\forall X_1,\ldots,X_n)$ $(\exists Y)(\forall x)(x\varepsilon Y\equiv\Phi(x,X_1,\ldots,X_n))$.

A3 (existence of sets) $\mathrm{Set}(0)\&(\forall x,y)\ \mathrm{Set}(x\cup\{y\})$

A4 (induction) For every <u>set</u>-formula Φ we accept the axiom $(\forall x,y)(\Phi(x)\rightarrow\Phi(x\cup\{y\}))\rightarrow(\forall x)\Phi(x)$.

Using these axioms we are able to define, as usual, the class of all natural numbers $N=\{x;(\forall y\varepsilon x)(y\subseteq x)\&(\forall y,z\varepsilon x)(y\varepsilon x\vee y= =x\vee x\varepsilon y)\&(\forall u\subseteq x)(u\neq 0\rightarrow(\exists q\varepsilon u)(q\cap u=0))\}$. The class FN of all finite (or "accessible") natural numbers is defined by $FN=\{\alpha\varepsilon N;\ (\forall X\subseteq\alpha)\ \mathrm{Set}(X)\}$. We formalize the intuitive notion "accessible natural number" by the notion "finite natural number"; we use the adjective "finite" in the alternative set theory only in such a sense, since in AST all sets are finite from the point of view of Cantor's set theory. Let us emphasize that both FN and N are models of Peano arithmetic and the very existence of such models is a base of possibility of two different views, e.g. comparison of discrete and continuous.

An ordering of A is called well-ordering if each non-empty sub-<u>class</u> of A has the first element; $X\approx Y$ ($X\preccurlyeq Y$ resp.) denotes that there is a one-one mapping of X onto (into resp.) Y and a class X is called countable (finite resp.) iff $X\approx FN$ ($X\prec FN$ resp.).

A5 (prolongation axiom) Every countable function can be extended in a set function, i.e. $(\forall F)((\mathrm{Fnc}(F)\&F\approx FN)\rightarrow(\exists f)(\mathrm{Fnc}(f)\&F\subseteq f))$.

A6 (axiom of choice) The universal class V (class of all sets) can be well-ordered.

A7 (axiom of cardinalities) Every two uncountable classes are equivalent, i.e. $(\forall X,Y)(X \preccurlyeq FN \vee Y \preccurlyeq FN \vee X \approx Y)$.

A8 (regularity) For every set-formula Φ we accept the axiom $(\exists x)\Phi(x) \rightarrow (\exists x)(\Phi(x) \& (\forall y \varepsilon x)(\neg \Phi(y)))$.

The axioms A1,A3,A4 and A8 are equivalent to ZF theory of finite sets (ZF_{Fin}). Our first thesis(cf.2.)is expressed by A4; the axiom A5 expresses both second and third thesis (since $A5 \rightarrow (\exists x)(\exists X)(X \subseteq x \& \neg Set(X)))$ and the last thesis is formalized by A7.

Let us note that, quite precisely said, we require the axiom A4 for all formal formulas the length of which is a finite natural number (see [S1]).

(2) Rational and real numbers ([**V**]).

As usual we can define the class of rational numbers (RN) as the quotient field of integers $(N \cup \{-\alpha;\alpha \varepsilon N\})$ and for $x,y \varepsilon RN$ we define $x \doteq y \equiv (\forall n \varepsilon FN)(|x-y|<1/n \vee (n<x \& n<y) \vee (x<-n \& \& y<-n))$. Rational numbers x,y are said to determine the same real number iff $x \doteq y$; a rational number x is called infinitely small iff $x \doteq 0$.

Real numbers are viewed as appropriate approximations of rationals, useful for computations. For example, there is no rational x such that $x^2=2$, but there is a rational x such that $x^2 \doteq 2$ and consequently a real x with $x^2=2$. Computations with real numbers are specific computations with rationals characterized by the fact that infinitely small numbers are disregarded. This is how real numbers are used in all applications.

(3) Codable systems ([V])

A system of classes $\mathfrak{M}$ is called codable iff there are K,S such that $X \varepsilon \mathfrak{M} \equiv (\exists x \varepsilon K) X=S''\{x\}$. Codable systems can be grasped in AST. By prolongation axiom, e.g. the system of all countable classes is codable. For every set x the system of all <u>subsets</u> P(x) is a set, but if x is an infinite set then the system of all <u>subclasses</u> of x is not codable (it is very large).

(4) Languages and some properties of classes ([V]).

We are able to define the class L of all formal formulas (with ε and =) as usual; moreover we put $FL=L\cap FN$ and elements of FL are called finite formulas. L_A (FL_A resp.) denotes the class of formulas (finite formulas resp.) in which parameters from A are admitted.

Finite set-formulas play a more important role than arbitrary ones, since, for example, the satisfaction Sat can be uniquely defined only for set-formulas from FL_V. The classes of the form Sat" $\{\varphi\}$ are called set-theoretically definable. A class X is called Π-class (σ-class resp.) iff there is a sequence $\{X_n;\ n\varepsilon FN\}$ of set-theoretically definable classes with $X=\bigcap\{X_n;\ n\varepsilon FN\}$ ($X=\bigcup\{X_n;\ n\varepsilon FN\}$ resp.). We can prove that a class is both Π and σ iff it is set-theoretically definable. The system of all Π-classes and σ-classes is codable.

A class X is called revealed iff for each countable $Y\subseteq X$ there is a set u with $Y\subseteq u\subseteq X$; a class X is called fully revealed iff for every normal formula $\varphi(z,Z)$ of the language FL, the class $\{x;\ \varphi(x,X)\}$ is revealed. Each set-theoretically definable class is fully revealed.

Theorem. If $\{X_n;\ n\varepsilon FN\}$ is a descending sequence of revealed non-empty classes then $\bigcap\{X_n;\ n\varepsilon FN\}$ is a revealed non-empty class and $dom(X)=\bigcap\{dom(X_n);\ n\varepsilon FN\}$.

(5) A model of Cantor's set theory ([V]).

Theorem. There is a class A satisfying the following:
(a) $(\forall x,y\varepsilon A)(x=y \text{ iff } x\cap A=y\cap A)$;
(b) $(\forall x\varepsilon A)(x\cap A$ is at most countable);
(c) $(\forall X\subset A)(X$ at most countable $\rightarrow(\exists x\varepsilon A)(X=x\cap A))$;
(d) $(\forall x\subseteq A)(x\varepsilon A)$.

(6) Indiscernibility equivalence ([V]).

An equivalence $\doteq$ is called an indescernibility equivalence iff it is a Π-class such that for each infinite set u there are

$x,y\in u$ with $x\neq y \& y\doteq y$.

The reasons why we require that an indiscernibility equivalence is an intersection of countably many set-theoretically definable classes were mentioned in 2.(7). The second requirement was stated since it is in the nature of an infinite set that we cannot see simultaneously its whole and each of its elements individually. Of course, the shape of a set depends on the methods of observation, and therefore on the corresponding indiscernibility equivalence. For example, if we observe a table (consisting of atoms) we perceive a shape entirely different from the shape perceived by an observer the size of an atom.

Indiscernibility equivalences play an important role in the alternative set theory. From the topological point of view we can grasp classical compact metric spaces in this way. Let us mention that the relation $\doteq$ we dealt with in 4.(2) is an indiscernibility equivalence.

Let us fix an indiscernibility equivalence $\doteq$. The figure of X (w.r.t. $\doteq$) is defined by $Fig(X)=\{y;(\exists x\in X)(x\doteq y)\}$; a class X is called a figure iff $X=Fig(X)$. The closure of X is defined by $\overline{X}=\{x; \neg(\exists Z)(Z$ is set-theoretically definable & $Fig(\{x\})\subseteq Z \& Fig(x)\cap Z=0)\}$. The closure operation has the usual topological properties, e.g. $X\subseteq\overline{X}$, $\overline{X\cup Y}=\overline{X}\cup\overline{Y}$ and $\overline{X}=\overline{\overline{X}}$; X is called closed iff $\overline{X}=X$.

Theorem. If X is set-theoretically definable then $Fig(X)$ is a π-class.

Theorem. If X is a figure then the following are equivalent: (a) X is a figure of a set; (b) X is a π-class; (c) X is revealed; (d) X is closed.

Theorem. There is at most countable class Z such that for each closed figure X we have $X=\bigcap\{Fig(u);\ u\in Z \& X\subseteq Fig(u)\}$.

(7) Motion ([V]).

Let $\vartheta\in N$ and let $\doteq$ be an indiscernibility equivalence. A function d is a motion of a point (of a set resp.) in the time ϑ iff $dom(d)=\vartheta+1$ and, for each $\alpha<\vartheta$ we have $d(\alpha)\doteq d(\alpha+1)$ ($Fig(d(\alpha))=Fig(d(\alpha+1))$ resp.).

The following theorem shows how it is possible to imagine a motion of a set as motions of its individual points.

Theorem. Let d be a motion of a set in the time ϑ such that $(\forall z)(\forall\alpha<\vartheta)(\mathrm{Fig}(\{z\})\cap d(\alpha)\approx \mathrm{Fig}(\{z\})\cap d(\alpha+1))$ then there is a system T of motions of points such that

(a) $(\forall\alpha<\vartheta)(d(\alpha)=\{f(\alpha);\ f\varepsilon T\})$

(b) $(\forall\alpha<\vartheta)(\forall f,g\varepsilon T)(f\neq g\rightarrow f(\alpha)\neq g(\alpha))$

(c) $(\forall\alpha<\vartheta)(\forall u)(\mathrm{Set}(\cup\{f\varepsilon T;f(\alpha)\varepsilon u\}))$.

The first statement implies that every point of the given set d(0) has its motion in T; (b) conveys that two different motions of points in T cannot go through one point. The third statement expresses the fact that the system T determines, moreover, the motion of every subset of d(0).

(8) Similarities ([V]).

We define $x \overset{\circ}{=} y$ iff for each set-formula $\varphi(z)$ of FL we have $\varphi(x)\equiv\varphi(y)$. This equivalence plays an important role in many branches of mathematics in the alternative set theory.

Theorem. $\overset{\circ}{=}$ is an indiscernibility equivalence (having a series of further interesting topological properties).

Theorem. Y is a clopen figure w.r.t. $\overset{\circ}{=}$ iff there is a set-formula φ of the language FL(!) such that $y=\{x;\varphi(x)\}$.

A set y is definable (definable using parameters from a class X resp.) iff there is a set-formula φ of the language FL(FL_X resp.) such that the formulas $(\exists!z)\varphi(z)$ and $\varphi(y)$ hold. We use the notation Def (Def_X resp.) for the class of all definable sets (definable using parameters from the class X resp.). The class Def is countable.

Theorem. The closure of Def w.r.t. $\overset{\circ}{=}$ is the universal class.

Theorem. If there is a formula $\varphi(z)$ of FL (not necessarily set-formula) with $(\exists!z)\varphi(z)\,\&\,\varphi(y)$ then $y\varepsilon \mathrm{Def}$.

Theorem. If $\varphi(z)$ is a set-formula of FL then the class $\{x;\varphi(x)\}$ is either empty or has a definable element.

A function F is called a similarity iff for each set-formula φ of FL and each $y_1,\ldots,y_n \in \mathrm{dom}(F)$ we have $\varphi(y_1,\ldots,y_n) \equiv$

$\equiv \varphi(F(y_1)\ldots,F(y_n))$. A similarity whose domain is the universal class is called endomorphism and endomorphism whose range is the universal class, too, is said to be automorphism.

Theorem. If F is a similarity, F at most countable, then there is an automorphism G such that $F \subseteq G$.

Theorem. For each y, y is definable iff $F(y)=y$ for each automorphism F.

A system $\mathfrak{M} \neq 0$ of set-theoretically definable classes is an ultrafilter (on the system of all set-theoretically definable classes) iff (a) $(\forall X,Y \in \mathfrak{M})(X \cap Y \in \mathfrak{M})$; (b) If Y is a set theoretically definable class then $(X \in \mathfrak{M} \,\&\, X \subseteq Y) \rightarrow Y \in \mathfrak{M}$;
(c) For every set-theoretically definable class X we have $X \in \mathfrak{M} \equiv (-X) \notin \mathfrak{M}$.

If $\mathfrak{M}$ is an ultrafilter then F, $\mathfrak{M}$, d are coherent if $\{x;\varphi(x,x_1,\ldots,x_n)\} \in \mathfrak{M} \equiv \varphi(d,F(x_1),\ldots,F(x_n))$ for each set-formula φ of FL and each $x_1,\ldots,x_n \in \mathrm{dom}(F)$.

Theorem. If F is a similarity which is at most countable and $\mathfrak{M}$ is an ultrafilter then there is a set d such that F, $\mathfrak{M}$, d are coherent.

Theorem. If F, $\mathfrak{M}$, d are coherent, F at most countable then there is an endomorphism F′ such that $F' \supseteq F$ and F′, $\mathfrak{M}$, d are coherent.

(9) Ultrafilters of sets ([S-V 4]).

An ultrafilter $\mathfrak{M}$ is ω-complete iff no element of $\mathfrak{M}$ is finite set and for each descending sequence $\{X_n;\ n \in FN\}$ of elements of $\mathfrak{M}$ there is an $X \in \mathfrak{M}$ with $X \subseteq \bigcap\{X_n;\ n \in FN\}$.

If we restrict ourselves to ultrafilters containing a set then the following two segments of N naturally characterize an ultrafilter $\mathfrak{M}$:

$\mu(\mathfrak{M}) = \{\alpha \in N; (\forall p)((p \hat{\lessgtr} \alpha \& Up \in \mathfrak{M}) \rightarrow p \cap \mathfrak{M} \neq 0)\}$
$\nu(\mathfrak{M}) = \{\alpha \in N; (\forall x \in \mathfrak{M})(\neg x \hat{\lessgtr} \alpha)\}$
where $x \hat{\lessgtr} y$ means that there is an one-one mapping of x into y which is a set.

Theorem. If $\mathfrak{M}$ is a nontrivial ultrafilter containing a set then
(a) $\alpha, \beta \in \mu(\mathfrak{M}) \rightarrow \alpha\beta \in \mu(\mathfrak{M})$
(b) $(\alpha \in \mu(\mathfrak{M}) \& \beta \in \nu(\mathfrak{M})) \rightarrow \alpha\beta \in \nu(\mathfrak{M})$
(c) $FN \subseteq \mu(\mathfrak{M}) \subseteq \nu(\mathfrak{M})$
(d) if $\nu(\mathfrak{M})$ is a π-class then $\mathfrak{M}$ is not ω-complete
(e) if $\mathfrak{M}$ is not ω-complete then $\mu(\mathfrak{M}) = FN$.

The following two results show that for every segments $R \subseteq S \subset N$ there are ultrafilters $\mathfrak{M}, \mathfrak{M}'$ such that $\mu(\mathfrak{M}) = \mu(\mathfrak{M}') = R \& \nu(\mathfrak{M}) = \nu(\mathfrak{M}') = S$ and $\mathfrak{M}$ is ω-complete and $\mathfrak{M}'$ is not ω-complete under the assumption that the existence of such ultrafilters is not excluded by the previous theorem.

Theorem. If a proper semiset S is a segment which is closed w.r.t. + then there is a nontrivial ultrafilter $\mathfrak{M}$ which is not ω-complete and such that $\nu(\mathfrak{M}) = S$.

Theorem. If $FN \subseteq R \subseteq S \subset N$ are two segments such that S is no π-class and such that the formula $(\forall \alpha \in R)(\forall \beta \in S)(\alpha\beta \in S) \& (\forall \alpha, \beta \in R)(\alpha\beta \in R)$ holds then there is an ω-complete ultrafilter $\mathfrak{M}$ with $\mu(\mathfrak{M}) = R \& \nu(\mathfrak{M}) = S$.

(10) Endomorphic universes([S-V1],[Ve 1],[Ve 2],[S-Ve]).

A class A is called an endomorphic universe iff there is an endomorphism F with rng(F)=A. An endomorphic universe is a copy of the universal class conveniently put into the universal class.

Theorem. The following properties of a class A are equivalent:
(a) A is an endomorphic universe:
(b) if $\varphi(Z_1,\dots,Z_n)$ is a normal formula of the language FL_A then for every $X_1,\dots,X_n$ the subclasses of A which are at most countable we have $\varphi(X_1,\dots,X_n) \equiv \varphi^A(X_1,\dots,X_n)$ where φ^A is the formula resulting from φ by the

restriction of all quantifiers binding sets variables to elements of A and all quantifiers binding class variables to subclasses of A;

(c) A satisfies the following two conditions:

(1) If $\varphi(z)$ is a set-formula of FL_A then we have

$$(\exists x)\varphi(x) \rightarrow (\exists x\varepsilon A)\varphi(x)$$

(2) For every countable $F\subseteq A$ there is $f\varepsilon A$ with $F\subseteq f$; (f,F being functions)

(d) if $\{\varphi_n(z);\ n\varepsilon FN\}$ is a sequence of set-formulas of FL_A then we have $(\exists x)(\forall n\varepsilon FN)\varphi_n(x) \rightarrow (\exists x\varepsilon A)(\forall n\varepsilon FN)\varphi_n(x)$.

The class Def (which is no endomorphic universe) can be obtained as the intersection of two endomorphic universes and even as the intersection of a descending sequence of endomorphic universes. On the other hand the intersection of at most countably many revealed endomorphic universes is an endomorphic universe. The union of any countable increasing sequence of endomorphic universes is no endomorphic universe.

Theorem. Let $\mathfrak{M}$ be a codable system of endomorphic universes such that for every $\mathfrak{N}\subseteq\mathfrak{M}$ which is at most countable there is $A\varepsilon\mathfrak{M}$ with $\bigcup\mathfrak{N}\subseteq A$. Then $\bigcup\mathfrak{M}$ is an endomorphic universe.

Theorem. If A and B are endomorphic universes then there is the smallest endomorphic universe a subclass of which is $A\cup B$.

Theorem. There is a transitive endomorphic universe which is the intersection of countably many sets.

Theorem. If A is an endomorphic universe then $\bigcup A$ is a revealed endomorphic universe.

For arbitrary class A and arbitrary set d we put

$$A[d]=\{f(d);\ d\varepsilon dom(f)\ \&\ f\varepsilon A\}.$$

Theorem. Let A be an endomorphic universe and let $d\varepsilon\bigcup A$. Then the class $A[d]$ is the smallest endomorphic universe subclass of which is the class $A\cup\{d\}$.

We define the reserve of X w.r.t. Y for each X,Y by $Rsv(X,Y)=\{x;\ Def_{X\cup\{x\}}\cap Y=0\}$.

Theorem. Let the pair $\langle X,Y\rangle$ (i.e. the class $X\times\{0\}\cup Y\times\{1\}$) be fully revealed. then the class $Rsv(X,Y)$ is revealed. Furthermore, if X,Y are σ-classes then $Rsv(X,Y)$ is revealed.

Theorem. If $Rsv(X,Y)$ is a revealed non-empty class then there is an endomorphic universe A such that $X\subseteq A$ and $Y\cap A=0$.

Theorem. Let $Rsv(X,\{y\})$ be a revealed class for each set y. Then the intersection of all endomorphic universes containing X is the class Def_X.

Theorem. There is an endomorphic universe A such that the initial segment $\{\alpha\in N;\ \alpha\subseteq A\}$ with the usual addition and multiplication is not model of Peano arithmetic.

Theorem. If $w\notin Def_w$ then there is an endomorphic universe A such that $w\subseteq A$ and $w\notin A$. Moreover each infinite set u has a subset w such that $w\notin Def_w$.

Theorem. There is an endomorphic universe A with

$$(\forall\gamma\in N-FN)(\exists\alpha,\beta\in\gamma-FN)(\alpha\in A\ \&\ \beta\notin A).$$

Theorem. Let x,y,d be sets, $Def_x\cap(y\cup\{d\})=0$, $d\in\bigcup Def_x$ and

$x,y\in Def_{(x\cup\{d\})}$.

Then there is an endomorphic universe A so that $x\subseteq A$, $y\cap A=0$ and $A[d]=V$.

Theorem. There is an endomorphic universe A such that the equivalence $\varphi(X_1,\dots,X_n)\equiv\varphi^A(X_1,\dots,X_n)$ holds even for any

$X_1,\dots,X_n$ semisets in the sense of A and any formula $\varphi\in FL_A$

in which only semisets are quantified.

(11) Similarities extendable to automorphism (Ve[2]).

Theorem. Let u be a set, H a countable class. Suppose that $F=H\cup Id\restriction(u\cup\{u\})$ is a similarity and $rng(F)=dom(F)$. (Id denotes the identity mapping). Then there is an automorphism G with $F\subseteq G$.

Theorem. Let a set-function f be a similarity. Suppose that there is a similarity $F\supseteq f$ such that $dom(F)$ is the smallest class T satisfying $T \supseteq dom(f)$ and $T \supseteq P(T)$ and for $u \subseteq dom(F)$ the equality $F(u)=F"u$ holds. (Such F is uniquely determined). Then there is an automorphism G with $f \subseteq G$.

(12) Revealments ([S-V 2])

A codable system of classes is called fully revealed iff there is its coding pair (cf.4.(3) which is fully revealed. If φ is a formula of FL_V and if $\mathfrak{M}$ is a codable system then $\varphi^{(\mathfrak{M})}$ denotes the formula resulting from φ by restriction of all quantifiers binding class variables to elements of $\mathfrak{M}$ (quantifiers binding set variables are let without change).

It is convenient to associate with every class a fully revealed class fulfilling the analogical properties as the original class. Hence we define that a class X is a revealment of a class Y iff X is a fully revealed class such that the formula $\varphi(X) \equiv \varphi(Y)$ holds for every normal formula $\varphi(Z)$ of FL. Analogically we say that a codable system $\mathfrak{M}$ is called a revealment of a codable system $\mathfrak{N}$ iff $\mathfrak{M}$ is fully revealed system of classes such that $\varphi^{(\mathfrak{M})} \equiv \varphi^{(\mathfrak{N})}$ holds for every φ of FL.

Theorem. Every codable system of classes has a revealment; in particual every class has a revealment.

Theorem. Let a codable system $\mathfrak{M}$ be a revealment of a codable system $\mathfrak{N}$. Then a codable system $\mathfrak{M}'$ is a revealment of $\mathfrak{N}$ iff there is an automorphism F with $\mathfrak{M}'=F"\mathfrak{M}$. In particular, if a class X is a revealment of a class Y then a class Z is a revealment of Y iff there is an automorphism F with $Z=F"X$.

Theorem. A codable system $\mathfrak{M}$ is fully revealed iff it fulfills the following conditions:

Rv_1 If $\varphi(z,Z_1,\ldots,Z_k)$ is a formula of FL_V and if $X_1,\ldots,X_k$ are elements of $\mathfrak{M}$ then the class $\{x;\ \varphi^{(\mathfrak{M})}(x,X_1,\ldots,X_k)\}$ is fully revealed.

Rv_2 If $\{\varphi_n(z,Z_1,\ldots,Z_k);\ n\varepsilon FN\}$ is a sequence of formulas of FL_V then we have $(\forall n\varepsilon FN)(\exists X\varepsilon \mathfrak{M})(\varphi_o(X,X_1,\ldots,X_{k_o})\ \&\ldots\&\ \varphi_n(X,X_1,\ldots,X_{k_n})) \rightarrow (\exists X\varepsilon \mathfrak{M})(\forall n\varepsilon FN)\varphi_n(X,X_1,\ldots,X_{k_n})$.

Theorem. Let $\varphi(Z)$ be a formula of FL and let a codable system $\mathfrak{M}$ be a revealment of a codable system $\mathfrak{N}$. Then the class $\{x;\varphi^{(\mathfrak{M})}(x)\}$ is a revealment of the class $\{x;\varphi^{(\mathfrak{N})}(x)\}$. In particular, if a class X is a revealment of a class Y then the class $\{x;\varphi(x,X)\}$ is a revealment of the class $\{x;\varphi(x,Y)\}$ under the assumption that $\varphi(z,Z)$ is a normal formula of FL.

Theorem. If $\overline{X}$ denotes the closure of X in the indiscernibility equivalence $\overset{\circ}{=}$(cf.4.(8))then $\overline{X} = \bigcup\{Y;\ Y$ is a revealment of $X\}$.

Theorem. A codable system of classes has exactly one revealment iff there is its coding pair of the form $\{x;\ \varphi(x)\}$ where φ is a set-formula of FL.

(13) Standard extension ([S-V 1]).

There are many different endomorphic universes in the alternative set theory. In many cases there is room enough for the natural extension of endomorphic universes. In this way we can imitate results obtained by Robinsons's nonstandard methods.

Let A be an endomorphic universe. An operation Ex defined for all sublcasses of A is called a standard extension on A iff for arbitrary normal formula $\varphi(Z_1,\ldots,Z_n)$ of FL_A and arbitrary $X_1,\ldots,X_n \subseteq A$ we have

$$\varphi^A(X_1,\ldots,X_n) \equiv \varphi(Ex(X_1),\ldots,Ex(X_n)).$$

In particular, we get that the Gödel's operations and Ex commute. It is possible to show that every endomorphic universe has at most one standard extension and that there are endomorphic universes having standard extension.

Theorem. The class Ex(X) is fully revealed for every $X \subseteq A$.

Theorem. The class Ex(X)-X is revealed for every $X \subseteq A$ and moreover if X is not finite then Ex(X)-X is non-empty.

(14) The axiom of reflection ([S-V 3])

This axiom can serve as an example of a statement which is intuitively true in the alternative set theory and which is not contained in the axiomatic system AST (since it is not necessary for many considerations).

This axiom asserts the existence of system of classes (called reflecting systems) which on one side are codable ("small") and which on the second side have the "same" property as the system of all classes.

We say that a system of classes $\mathfrak{M}$ is a reflecting system iff the following two conditions hold:

(a) for every formula $\varphi(Z_1,\ldots,Z_k)$ of FL_V and every

$X_1,\ldots,X_k \in \mathfrak{M}$ we have $\varphi(X_1,\ldots,X_k) \equiv \varphi^{(\mathfrak{M})}(X_1,\ldots,X_k)$

(b) if $\{X_n;\ n \in FN\} \subseteq \mathfrak{M}$ then there is a coding pair $\langle K,S \rangle \in \mathfrak{M}$ which codes $\{X_n;\ n \in FN\}$.

<u>Axiom of reflection</u>. For every class X there is a <u>codable</u> reflecting system $\mathfrak{M}$ with $X \in \mathfrak{M}$.

The axiom of reflection makes it possible to introduce new types of standard extensions (defined on reflecting systems) and in this way we are able to imitate, for example, enlargements of non-standard methods.

(15) Elimination of infinitely small quantities ([Č])

Let $\Phi(X)$ be a formula describing a property of parts of real numbers using infinitely small quantities or infinitely large numbers (i.e. $\Phi(X)$ is a formula in which only real numbers are quntified with one free variable X for parts of real numbers using the predicate "to be an infinitely large natural number"). In particular, we deal with all properties of real functions of n variables described in infinitesimal calculus of Leibniz type. Infinitely large natural numbers can be quantified on various places in Φ. There is a procedure of how to find a formula $\Psi(X)$ describing the same property in which we use only auxiliary variables for real and natural numbers and we do not use infinitely large and infinitely small quantities.

(16) Valuations ([M 1], [M 2], [M 3]).

Theorem. Let R be an equivalence on a. Then there exists a metric $h: a^2 \to RN(\geqslant 0)$ (where $RN(\geqslant 0)=\{x\geqslant 0;\ x\in RN\}$) so that $h(x,y)\doteq 0 \equiv xRy$ holds. More generally: let Sd^*_V be a revealment of the system of all set-theoretically definable classes and let $\overset{\circ}{=}$ be a π-equivalence on a class $A\in Sd^*_V$. Then there exists a metric $H\in Sd^*_V$ so that $H(x,y)\doteq 0 \equiv x\overset{\circ}{=}y$ holds but there is no set-theoretically definable metric with the same property.

Theorem. Let Q be a π-ideal (σ-ideal resp.) on P(a). Then there exists a mapping $h: P(a) \to RN(\geqslant 0)$ ($h: P(a) \to N$ resp.) so that $h(x)\doteq 0 \equiv x\in Q$ ($h(x)\in FN \equiv x\in Q$ resp.) and we have $h(x\cup y)=$ $=h(x)+h(y)$ and $x<y \to h(x)<h(y)$.

The results presented are special cases of general valuation theorem.

(17) Representations of models of PA and ZF.

The following results show how it is possible to represent isomorphically models of Peano arithmetic PA and models of ZF or ZF_{Fin} as substructures of the universal class (the second result can be essentially strengthened).

Theorem. Let $M \models PA$. There exists a substructure M^* of $\langle N,0,1,+\rangle$ and set-function f so that M and $\langle M^*, f\cap(M^*)^3\rangle$ are isomorphic.

Theorem. Let $M \models ZF$ of $M \models ZF_{Fin}$. Then there exists an extensional class A so that A contains all its finite subsets, the operations $\cup$, $\cap$ and $-$ are absolute for A and M and $\langle A,\in\rangle$ are isomorphic.

(18) Real and imaginary classes ([Č-V 1], [S-Ve]).

Our every observation is characterized by an indiscernibility equivalence. The classes, we observe on the horizon of our observing abilities, are exactly the figures in the mentioned indiscernibility equivalence. These considerations lead us to the following definitions:

A class is called real iff it is a figure in an indiscernibility equivalence. Classes which are not real are called imaginary.

Theorem. Let F be an automorphism. If X is a real class then F"X is real, too.

Theorem. If $\varphi(z,Z_1,\ldots,Z_n)$ is a formula of FL_V and $Y_1,\ldots,Y_n$ are real classes then the class $\{x;\ \varphi(x,Y_1,\ldots,Y_n)\}$ is real.

Theorem. The system of classes {F"X; F is an automorphism} is codable iff X is a real class.

Theorem. Let $\doteq$ be an indiscernibility equivalence and let X be its selector. Then X is either finite or imaginary. The only real automorphism is the identity.

Theorem. Endomorphic universe is imaginary iff it is no π-class.

Theorem. For every real function F there is a sequence $\{F_n;\ n\in FN\}$ of set-theoretically definable functions such that $F\subseteq \bigcup\{F_n;\ n\in FN\}$.

Theorem. If $\alpha\notin FN$ then there is a real one-one mapping of α onto $2\cdot\alpha$ and on the other hand every one-one mapping of α onto α^2 is imaginary.

Theorem. Let X be a real class, $\alpha\in N$. Then there is either a set u with $u\subseteq X \,\&\, u\,\hat{\approx}\,\alpha$, or for each $\gamma\in N-FN$ there is a set $u\supseteq X$ with $u\,\hat{\approx}\,\alpha\cdot\gamma$.

Theorem. If I is a real class of indiscernibles which is no semiset then I is a class of strong indiscernibles. Moreover there is a class of indiscernibles which is a π-class and which is no semiset.

(19) Metamathematics of AST ([S 1], [S 2], [S 3], [Ve 3]).

Let KM_{Fin} (KM^{-} resp.) denote Kelley-Morse set theory in which the axiom of infinity is replaced by its negation (the power set axiom is omitted resp.).

It is possible to show that ZF_{Fin} is equivalent to the system of those axioms of AST which concern sets only and further that KM_{Fin} is equivalent to the theory which we get from AST replacing the prolongation axiom by the axiom guaranteeing that every subclass of a set is a set.

The results concerning interpretability of investigated theories can be drawn in the following diagram, where $\rightarrow$ denotes the existence of an interpretation of the first theory in the second one and $\not\rightarrow$ represents the non-existence of such interpretation.

$$\begin{array}{ccccc} KM_{Fin} & \overset{\longrightarrow}{\not\longleftarrow} & AST & \overset{\longrightarrow}{\not\longleftarrow} & ZF \\ & & \downarrow\uparrow & & \\ & & KM^{-} & & \end{array}$$

We see from the above diagram that AST is "strictly stronger" than KM_{Fin} in the sense of interpretability, but the position of these two theories changes if we investigate the provability of set-formulas. It can be shown that in AST there are provable exactly these set-formulas as in ZF_{Fin} and therefore one is able to prove in AST less set-formulas than in KM_{Fin}.

The axiomatic system AST is really extremely minimalized since every axiom of AST is independent on the other. Moreover, AST is not finitely axiomatizable.

Both theories AST + axiom of reflection and AST + $\neg$ axiom of reflection are consistent relatively to ZF.

Dealing in ZF (+CH) with models of AST we see that a model $M \models ZF_{Fin}$ can be extended to a model of AST by adding all its subsets as classes iff M is $\aleph_1$-saturated model of cardinality $\aleph_1$. If A⊨AST then the model of all A-sets (with A-membership relation) is recursively saturated.

REFERENCES

[B] Bolzano,B.: 1851,'Paradoxien des Unendlichen, Leipzig.

[Č 1] Čuda,K.: 1980,'An elimination of infinitely small quantities and infinitely large numbers', CMUC 21, 433-445.

[Č-V 1] Čuda,K. and P.Vopěnka: 1979,'Real and imaginary classes in the alternative set theory', CMUC 20, 639-653.

[M 1] Mlček,J.: 1979,'Approximations of σ-classes and π-classes', CMUC 20, 669-679.

[M 2] Mlček,J.: 1979,'Valuations of structures', CMUC 20, 681-695.

[M 3] Mlček,J.: 1981,'Monotonic valuations and valuations of triade of higher types', CMUC 22, 377-398.

[P-S] Pudlák,P. and A.Sochor: 'Models of the alternative set theory',to appear.

[R 1] Resl,M.: 1979,'On models in the alternative set theory', CMUC 20, 723-736.

[R-S] Resl,M. and A.Sochor: 1981, 'Provability in the alternative set theory', CMUC 22, 655-660.

[S 1] Sochor,A.: 1979, 'Metamathametics of the alternative set theory I', CMUC 20, 697-722.

[S 2] Sochor,A.: 1982,'Metamathematics of the alternative set theory II', CMUC 23, 55-79.

[S 3] Sochor,A.: 'Metamathematics of the alternative set theory III', to appear.

[So 1] Sochor,A.: 1979, 'Some philosophical problems and the alternative set theory', Teorie rozvoje vědy III, 105-120.

[So 2] Sochor,A.: 1977, 'Differential Calculus in the alternative set theory', Lecture Notes in Math. vol.619, 273-284.

[So 3] Sochor,A.: 1979,'Some remarks to the connection between the alternative set theory and nonstandard methods', Proceedings of Frege-Conference, Jena, 458-467.

[S-Ve] Sochor,A. and A.Vencovská: 1981, 'Indiscernibles in the alternative set theory', CMUC 22, 785-798.

[S-V 1] Sochor,A. and P.Vopěnka: 1979, 'Endomorphic universes and their standard extensions', CMUC 20, 605-629.

[S-V 2] Sochor, A. and P.Vopěnka: 1980, 'Revealments', CMUC 21, 97-118.

[S-V 3] Sochor,A. and P.Vopěnka: 1981, 'The axiom of reflection', CMUC 22, 87-111.

[S-V 4] Sochor, A. and P.Vopěnka: 1981, 'Ultrafilters of sets', CMUC 22, 689-699.

[Ve 1] Vencovská,A.: 1979, 'A method for constructing some endomorphic universes', CMUC 20, 655-668.

[Ve 2] Venkovská,A.: 1982, 'Constructione of endomorphic universes and similarities', CMUC 23, 557-577.

[Ve 3] Vencovská,A.: 'Independence of the axiom of choice in the alternative set theory',to appear.

[V] Vopěnka,P.: 1979, 'Mathematics in the alternative set theory',TEUBNER TEXTE, Leipzig.

[V'] Vopěnka,P.: 'Mathematics in the alternative set theory', Russian translation, to appear.

[V 1] Vopěnka,P.: 1979, 'The lattice of indiscernibility equivalences', CMUC 20, 631-638.

CMUC = Commentationes Mathematicae Universitatis Carolinae (Prague)

Mathematical Institute of Czechoslovak Acad. of Sci., Prague, Czechoslovakia

Settimo TERMINI

ASPECTS OF VAGUENESS AND SOME EPISTEMOLOGICAL PROBLEMS RELATED TO THEIR FORMALIZATION

> *The topic of vagueness is a very difficult one... it is,..., evident that we have an intuitive sympathy with Frege's view that a vague expression is one which is, to that extent, defective in sense: any account of vagueness which failed at least to explain this intuitive feeling would be inadequate. At the same time we have also a contrary intuition, that vagueness is an indispensable feature of our language, that we could not operate with language as we do unless many of its expressions displayed this feature: and it is equally a requirement on any satisfactory account of vagueness that it should at least explain our having this intuition also.*
>
> *Michael DUMMETT*

1. Introduction.

Vagueness, Approximation, Error, Imprecision, Accuracy etc. all share the same "family resemblance". The discussion in this paper will mainly concern vagueness although sometimes one or another of the other notions will show up.

This wandering through different notions will be permitted by the natural flexibility of the language which allows one to cross the dividing lines between them. The final aim of this meandering being, however, to use this flexibility to put in focus better the problem of vagueness and not to remain satisfied with vague descriptions.

The central role that imprecision plays in science is well

H. J. Skala, S. Termini, and E. Trillas (eds.), Aspects of Vagueness, 205–230.

known. For instance Ludwig (1981) writes: *"One of the most fundamental facts of experimental physics a student has to learn at the beginning of his studies is that no measurement is precise.(...) A theoretical physicist will realize that there is another field of imprecision. He will realize that every mathematical theory used as a picture of reality cannot be regarded as a precise picture. He will realize that there is no difference in principle between a so-called exact theory and an approximation picture of reality. What is the meaning of approximation and imprecision in this field?"*

The revisability of Science is presented by Parrat[1](1961) in this way: *"Every* fact *in Science, every* law of nature *as devised from observations is intrinsically "open-ended", i.e. contains some uncertainty and is subject to future improvement" (page 1).* He however stresses that *"the feature that distinguishes scientific knowledge is not only that there is a clear recognition of uncertainty but that the degree of uncertainty can usually be rather well determined" (page 56).*

This attitude is not exclusive to physics. Similar attitudes can be found also in other disciplines. In one of the first papers on the subject, for instance, John von Neumann (1956) introduces the role of error in the theory of automata thus: *"The subject-matter is the role of error in logics, or in the physical implementation of logics - in automata synthesis. Error is viewed, therefore, not as an extraneous and misdirected or misdirecting accident, but as an essential part of the process under consideration - its importance in the synthesis of automata being fully comparable to that of the factor which is normally considered, the intended and correct logical structure" (page 43).*

It is more difficult to find in the scientific literature such an explicit reference to vagueness; it, however, can be assumed to be a unifying notion with respect to the different aspects of uncertainty and imprecision in a scientific theory. We shall take as a basis for the subsequent discussion some considerations of Karl Popper on the problem precision/imprecision in science, which incidentally turns out to be very similar to a comment of Aristotle.[2] What both Popper and Aristotle essentially suggest[3] is that *"precision for its own sake"* can be more dangerous than enlightening and that it is the problem itself which must determine the level of precision required for its satisfactory solution. Precision, then, does not exist *in abstracto;* least of all, it can be mechanically determined on the basis of the technical tools which are used.

Another question which apparently seems related to the previous problem is the dichothomy between "hard" and "soft" sciences. Is it possible to relate such a division to the fact that the problems of the "hard" sciences require a high level of precision while those of the "soft" sciences require low levels of precision for their solution? The answer is definitely in the negative. The challenging aspect of Popper's formulation of the problem is that it is not useful to speak of precision abstractly; the same division of sciences into "hard" and "soft" seems questionable.

To clarify what our attitude towards the problem of vagueness is, let us make a brief digression on another related question. The classification of sciences into hard and soft or of problems into precise and fuzzy is sometimes defended on the basis that such a division would capture a natural distinction between those sciences and problems which are concerned with very strong idealizations and those which represent more faithfully what actually happens in the real world. This position would then eventually consider precision and vagueness as attributes of the world.

This attitude is, in my view, unsatisfactory at least for the two following reasons. First it would force one to add new ontological committments different from those usually made in the scientific tradition. Secondly there is the fundamental objection that it is also hard to make sense at all of this same notion; it is in fact difficult not to agree with Michael Dummett that *"the notion that things might actually* be *vague, as well as being vaguely described, is not properly intelligible".*[4] If an hypothesis of this kind would be necessarily required to admit a general notion of vagueness into the scientific theories, one could rightly understand why "classical sciences" renounced attempts to deal with vagueness in general and confronted only particular aspects of vagueness in specific ways (examples: errors in empirical measurements; uncertainty relations in quantum theory; thermal and quantum fluctuations, etc.).

The acceptance of such an hypothesis would impose a reassessment of all the traditional construction of the scientific enterprise. This however seems not necessary since one can follow the position of Russell (1923) according to which "*vagueness or precision are characteristics which can only belong to a representation... they have to do with the relation between a representation and what it represents*".[5]

The overall view emerging from these considerations is that the general notion of vagueness can and must play a central role in scientific theories only under the two following conditions: a) It must be a unifying concept with respect to all the various aspects of "lack of precision" present in scientific theories; b) it must be compatible with the usual general presuppositions of science. Finally it should also be as innovative as possible with respect to the specific theories, approaches and traditional foundations (as a consequence of taking point a) seriously).

Let us stress that the whole paper is nothing more than a long memorandum for future research or, equivalently, a partial summary of an (epistemo)logical analysis to be made on all the material which has accumulated in recent years on the problem of vagueness. The following Section will be devoted to a review of some technical proposals which have been put forward for treating vague notions. In the concluding Section we shall briefly touch some conceptual points.

2. A Brief Analysis of Some Technical Proposals.

In the discussion which follows we shall not attempt to give a complete picture of all the ways in which the problem of vagueness has been tackled. We shall review only some approaches: their common starting point is the acknowledgment that some notions, terms, predicates, etc. which one would not readily give up do not behave "classically" (interpreting here the term "classical" in a very informal way, roughly as similar to the ideal concepts of traditional mathematics). Briefly the relevant "nonclassical" feature which one wishes to account for is a sort of unavoidable lack of precise boundaries for these concepts. One then aims at a theory providing the tools to deal with these concepts in the study of those systems in which vagueness is either ineliminable or eliminable at the price of an intolerable increase in their complexity. Moreover a theory of vagueness, whatever the latter might be, should avoid the emergence of "sorites"-like paradoxical situations.

In the foundations of physics the emergence of vague predicates is related to the acknowledgment that *"many elementary experiments are dispersive:. i.e., they have the property that a repetition of the experiment does not necessarily produce a repetition of the outcome"* (Giles, 1976). If one does not want to give up interpreting *"an expression to which there corresponds a definite elementary experiment"* as a prime proposition, then one is forced to accept the existence of *dispersive prime propositions*. A situation of this type cannot always be accounted for by the theory of probability and therefore can be considered as pertaining to the general topic of the presence of vagueness in science. Another aspect of physics (or, for what is of interest here, of any science or any empirical activity) which also belongs to this topic, is the problem of empirical measurements.

All these aspects, as well as others not mentioned at all, are obviously interrelated. It is not however completely clear which are the specific, *necessary* relations existing among the various problems involved. Let us add to the presence of borderline cases and the existence of dispersive prime propositions already mentioned the following ones: a) *the (in)coherence of natural language;* b) *the nature of mathematical truth (with respect to vague predicates);* c) *the problem discrete/continuum;* d) *the relationship between mathematics, empirical observations and reality, etc.*

A careful consideration of these aspects would immediately lead to a crossroad of many difficult conceptual problems. An implicit assumption underlying some, but not all, of the approaches that we shall consider is the following: a way to avoid passing through the "Caudin Forks" of the philosophical analyses related to vagueness/mathematics/science, could be just to develop formal tools able to cope with situations in which borderline cases or dispersive propositions are present, enabling one to solve all the relevant problems of a scientific type which arise.

Such a choice is, of course, restrictive and, in a certain sense, also arbitrary. We believe however that it can be a meaningful and useful starting point if one analyzes seriously all the consequences to which it leads. Among the merits of such an attitude let us mention, moreover, the possibility of providing theories which are relatively simple from a technical point of view and are, at the same time, very intuitive and appealing, conceptually. Another advantage of simple beginnings is that the problems involved can be faced and treated in their generality and not under the pressure of specific constraints of the particular domain of the phenomena considered. A general approach of this type could be useful also in the foundations of physics, where the alleged vagueness of some predicates has been studied almost always having in mind the specific features of the quantum domain.

A two-fold division is useful for the subsequent discussion. In the two following subsections we shall discuss respectively the "metrical" developments of Scott's version of Łukasiewicz logic and the modelling of vague predicates or of dispersive prime propositions by means of fuzzy sets, i.e. generalized (set-theoretic) characteristic functions.

2.1. A metrical approach

Dana Scott (1974), after expressing some reserves on the import of many-valued logic on the basis that a conceptual justification of the intermediate values is not immediately available, proposes to interpret this logic as a non probabilistic "logic of error", whose truth values would then measure the degree of error within which a certain statement is true. M. Katz (1981 and 1982) has developed this idea by providing a detailed study of two real-valued logics; the first, called logic of inexactness is essentially the infinite valued Łukasiewicz logic interpreted *à la* Scott: *"within this logic, which we want to use as a framework for science, incorporating the unavoidable errors of measurement, truth-values are thought of as representing degrees of error"* (Katz, 1981). The second logic, called logic of approximation, is a proper extension of the logic of inexactness and *"is intended to capture mathematically the idea of approaching truth as near as one wishes"*. A technical feature of these two systems is an axiomatization of a real-valued equality which induces metrics on the domains of appropriate structures: equality is then not two-valued, but the truth-value of the assertion of the equality of two objects becomes a function of some kind of "distance" between the objects. What is the relevance of these systems to the problems of vagueness which we are examining? The first of course, is that inexactness and approximation belong to the same family of notions and then any formal proposal for treating them will be, at least indirectly, relevant also for the general problems of vagueness. A second reason is given by the same interpretation of the truth-values as errors. This sounds very appealing. Even if we do not know really what vagueness is, on the premises of avoiding as far as possible preliminary analyses of a philosophical type, the idea of looking at borderline cases or at dispersive propositions through the lens of "error" sounds interesting whatever is the referent with respect to which one has committed the error. After all, the same standard Łukasiewicz infinite valued logic has been used directly also in Applied Engineering to get rid of the rigidity of the all-or-none concepts of classical approaches. Both the new interpretation of the truth-values and the introduction of a notion of real valued equality seem then favourable for formally treating vagueness through Łukasiewicz logic.

A consequence of this, however, is that one is naturally led to consider as satisfactory a way of technically tackling the problems of vagueness and approximation which essentially measures how far a certain vague or approximate description is "distant" from a corresponding description which is error-free and is taken as an ideal model. Now it may certainly seem natural to evaluate the distance between the two descriptions; however, only if this error-free ideal model (corresponding to the vague or approximate description) is assumed to exist. There are, then, some implicit assumptions which are made and which it will be better to examine in detail. Let us consider the cases of approximation and vagueness separately.

Approximation: One assumes that, a) there is something which one is approximating and b) there is a uniform way of approaching this something (as long as it is possible to define a "distance").
The first assumption is natural and fits in well with the intuitive idea of approximation. The second assumption, notwithstanding its innocent appearance, clashes with a result of Miller, which we shall discuss shortly, if we allow the term "approximation" to be interpreted also in terms of "numerical accuracy".

Vagueness: Also here two similar assumptions are made: a) that there is something to approach and b) that there is a uniform way of approaching it. What changes is the nature of this something: it must be a sharpening of our vague description which, according to the other assumption, one can sharpen uniformly. The reference to Miller's result is here less cogent and direct than in the previous case, however, one is left wondering whether this way of putting things is the right one.

To look at the problem in this way means to consider vagueness eventually only as a frill since once the (true) sharpened version of a vague term is known there are no reasons for not giving up the vague terms and considering only their corresponding sharpened versions.
The difficulty seems to reside just in the fact that there seem to be no natural ways of sharpening vague predicates.

Let us stress that the previous comments do not undermine the usefulness of these theoretical efforts and their possible practical relevance and effectiveness (See, for instance, Katz, 1981, § 3), for a useful overview of "Applications to Scientific Inquiry"). What we are questioning is whether these proposals can be considered "natural" formal explicata of the informal notions of "approximation" and "vagueness".

Let us consider another point: the fact of developing according to the usual standards of mathematics is certainly a merit of these theoretical efforts. The straightforward way in which they borrow and use the metrical concepts which are needed is, however, a handicap for using the proposal as an explicatum of vagueness in general. The price to be paid in fact as we argued previously, would be the assumption that vagueness has to be ascribed essentially to a lack of information, that is, to an incompleteness in the description. It is then clear that this approach can work perfectly in, and is perfectly adequate to, all those situations in which vagueness was misleadingly attributed to situations of an essentially classical nature even if incompletely described, but it seems to give no essential contribution to the clarification of the nature of vagueness and its innovative character.

A by-product of many classical epistemological analysis of vagueness (and also of our brief comments) is their stressing exactly its innovative character also with respect to the traditional approach of mathematics. It seems that to deal deeply with vague predicates points to the need of reformulating traditional ways of looking at mathematics. To conclude let us summarize the previous discussion by saying that if we must take vagueness (and, in a sense, also approximation) seriously it seems restrictive to confine ourselves, from the beginning, to formalizing it through classical mathematical notions *tout court* (in particular, by means of metrical notions) since, as we have suggested, doing so imposes very strong conceptual assumptions on the notion of vagueness, which destroy its apparent innovativeness without providing convincing reasons for its non-existence.

2.2. Fuzzy sets, Many-valued sets and Setts.

In this subsection we shall discuss briefly the approach to the problem of vagueness underlying the theory of fuzzy sets of Zadeh and the related proposals of Klaua and Giles. We refer to the well known literature for specific reference to standard Zadeh's theory. An overview of Klaua's work and its relationship to Zadeh's theory can be found in Gottwald's paper in this volume, while the general idea of Giles' approach can be found in his 1976 paper.

This approach starts from a very challenging idea and with an ambitious program. The idea is that in many situations one should consider as a point of departure the information as it is actually provided by a measurement, an observation, or the like. The associated predicates, terms or propositions must not then be preprocessed in order to transform them to fit the classical standards. The basic idea behind this approach is therefore that one has to model directly the intuition of a dispersive prime proposition or of the existence of borderline cases.

A very simple and natural way of doing so is through a generalization of the notion of characteristic function of classical set theory obtained by extending its range from $\{0,1\}$(when 0 means "not-belongs" and 1 means "belong") to the interval $[0,1]$ of the real line (or to a suitable lattice, in other formulation), thus providing set-theoretical ways of taking into account borderline cases and of mapping directly the dispersive prime propositions. The ambition of this approach is that the notion of a generalized characteristic function could be also a starting point for generalizing the concept of set, and capturing at this level, the innovative character of vagueness. We must admit that this second approach seems to face the problem of vagueness in a more radical way. We shall not review successes, failures and formal developments of the theory for which we refer the reader to the literature (for instance, Goguen (1974), Gaines (1976), Pavelka (1979), Zadeh (1975), Katz (1981) (§ 3c), Giles (1976) Gottwald (1983)). Instead we shall take into account and discuss briefly here a recent result obtained by Pultr in a paper reprinted in this volume which indirectly sheds light on the problem of the relationship between the notion of vagueness and mathematics. Pultr observes that a notion of approximate equality is an essential ingredient of any theory of vagueness which the theory of fuzzy sets claims to be. It seems then natural to enrich the theory with a suitable notion of fuzzy equivalence and study the theory so obtained. Surprisingly Pultr shows that *"roughly speaking, fuzzy sets endowed with fuzzy-equivalences can be adequately represented by means of the fuzzy-equivalences alone. In the representation the fuzzy sets appear as pairs (X',X") of crisp subsets of universa with a kind of distance (resulting from the fuzzy-equalities) and one operates with them crisply"*. We refer the reader to Pultr's paper for more specific information and technical details.

For evaluating the consequences of this result, which are relevant to our problem we select some distinct questions:

1) How natural is the fuzzy equivalence introduced by Pultr?

2) What is the import of this result on the theory of fuzzy sets in general?

3) What is the import of this result on using the theory of fuzzy sets as a theory of vagueness?

Let us discuss the various questions separately.

1) There is no doubt that the fuzzy equivalence introduced by Pultr has a distinct metrical flavour. So it may seem not unexpected, after all, that a representation result of metrical type is obtained. (This comment bears only on the conceptual unexpectedness, however, and has nothing to do either with the technical difficulty of actually proving the theorem or the mathematical deepness of the result). This fuzzy equivalence however, captures some intuitions in a very natural way. For instance, two intuitive properties that can be required of

a notion of imperfect equality can be written as:

1) $x =_a y$ and $y =_b z$ imply $x =_{ab} z$

2) $x \in_a X$ and $y =_b x$ imply $y \in_{ab} X$

with the obvious meanings of $=_a$ (equal to, with degree a) and $\in_a$ (belongs to, with degree a).

These requirements can be immediately accounted for by the metrical structure proposed by Pultr. So no violence whatever seems to be done to the intuitive requirements of the theory. Let us now pass to the second question.

2) The relevance of this theorem for the theory of fuzzy sets is, in my view, outstanding. It, in fact, determines in a rigorous way the relationship existing between fuzzy models and classical models. On the one hand, it establishes that the type of imprecision captured by the theory of fuzzy sets through the notion of generalized characteristic function can be rigorously represented by means of classical structures. On the other hand fuzzy models based on, or using the theory of fuzzy sets can be seen as rigorous representations of models of a metrical type. It is true that the theorem says, in principle, that there is no theoretical gain in working with the theory of fuzzy sets since all that can be done with this theory could also be done with classical structures of a metrical type, but this has a validity only as a matter of principle. The original representation and idea of a fuzzy set is so intuitively simple and appealing that his heuristic role cannot be substituted easily except through ideas equally intuitively simple and appealing. If fuzzy sets represent the information immediately available and therefore the primary elements empirically available (let us not forget that this was just the starting point of the approach we are now analyzing) then the previous theorem can be used the other way around to establish the kind of distances adequate to express the fuzzy equality. Pultr's theorem then, in my view, should be particularly appreciated by those interested in applications since it provides a way of translating into a classical rigorous language of mathematics the incomplete information that is immediately available and which can usefully be modelled directly through the theory of fuzzy sets. The reader may

have been surprised by the positive attitude towards the use of metrical notions shown in this discussion, especially when compared with the somewhat bitter attitude manifested in the preceding subsection, but, in fact, we have not touched at all the implications of a conceptual type, which is the specific content of question 3.

3) What is the impact of this theorem if one wishes to use the theory of fuzzy sets as a theory of vagueness? One could simply refer to the considerations of the previous subsections and conclude that it is devastating. The approach considered in this subsection, which raised hopes of coping satisfactorily with the problem of vagueness on the basis of a seeming major radicality, fell short of their expectations. Notwithstanding its appearance, this approach is not so radical as the radicality which the problem requires. However, it is worthwhile to look at the question more carefully. One could rightly observe that, after all, Pultr's theorem is not so compelling as it appears since the fuzzy equivalence introduced, even if natural, is itself based on metrical concepts. However it seems very unlikely that, even by introducing quite different fuzzy equivalences, we would be forced to change substantially the previous conclusions. What could happen at the most would be to acknowledge the possibility of dispensing with metrical structures. In fact, upon reflection, one realizes that Pultr introduced the sin and the temptation into the Garden of Eden of fuzzy sets not so much through the use of metrical structures but more by insisting on the importance of defining a fuzzy equivalence and looking for global representations of the enriched theory.

Pultr's representation theorem is based on a structural idea that, in a sense, goes beyond the specific fact of actually using metrical concepts and tools. This structural idea appears to be of no less epistemological interest than the theorem itself. As Pultr writes in the quotation above, through his theorem, a fuzzy set can be represented "crisply" by means of a couple of crisp subsets (one representing essentially a kernel of objects to which a certain property can be undisputedly attributed, the other the domain of definition of the fuzzy set associated with the same property) and a subsidiary notion defined in terms of the fuzzy equivalence (which in the specific case turns out to be a distance). Epistemologically the theorem stresses the usefullness of obtaining classical representations enriched by suitable additional structures determined by the notion of imperfect equality. The metrical notions involved and used are essential technically, since they allow one to prove the theorem, but not epistemologically.

The theorem then suggests looking for non metrical representations of a classical type, should it turn out that different fuzzy equivalences capture the intuitive notion even better than that used by Pultr. Despite this, the representation would still remain of a "classical" type, showing then the strong resistance of the approach to accept adequately the request for innovativeness made by the notion of vagueness.

This, however, is only an additional positive feature of an epistemological type of this theorem. Its "hard" merit, of course, resides in the fact that the given representation is proved rigorously and on the basis of assumptions compatible with the intuitive requirements.
All the previous discussion seems to us to point to a conclusion which is in contrast to a widespread feeling according to which the conceptual aspects of a theory of vagueness are clear and the actual problems reside in the technical difficulties of treating and developing the main points adequately. Our feeling however is that traditional mathematics is able to deal with the problems of vagueness innovatively less at a conceptual level than at a technical level. Technically we can deal with this but the problem left concerns the fact that the only sound interpretation of this treatment is not innovative. The technical results seem to be able to capture the notion of vagueness if this is interpreted in a sense that is very traditional, conceptually more than technically. The basis for doing this is, in fact, the construction of "idealizations" of the vague notions involved; once granted this, there is nothing unnatural in seeing how far these idealizations are from the vague notions from which we started and which definitely represent the actual central kernel of our problems.

This approach has a non negligible interest for the notion of vagueness in general only in a different sense, namely for measuring how reliable certain idealizations are. But the leading role on the stage must be played by vague notions, not by their idealizations; otherwise there is no hope of taking into account "nuances" of vagueness different from the "traditional" one of *"incompletely described"*. We are thinking of the intuitive aspects represented by the notions of *being discrepant, being partly incompatible, being not fully expres-*

sible within, etc., notions that one would desire to see embodied in the theory and which seem to clash with some general assumption of traditional mathematics. Perhaps one should sharpen a preceding conclusion by saying that classical, traditional, mathematics is able to deal technically with the problems of vagueness, however only for those aspects which are compatible with the traditional ontology of mathematics.

The radicality of the approach of fuzzy sets seems, then, to have been tempered by a natural conservatorism which leads us to accept too much of the received views on mathematics and on the way in which mathematics bears on the empirical world. Any innovation, however, must not question the level of rigour to which we are accustomed in traditional mathematics if we wish to preserve the dialogue between various approaches and disciplines which has always been a positive feature of the scientific discourse. Moreover relaxing rigour would destroy the possibility of both comparing the results obtained in the new setting with those of former theories and testing rigorously the new hypotheses, pushing them to their extreme consequences. As with all new innovative theories, a formal approach to vagueness should then be more demanding on rigour than traditional theories even if it should also be very sensitive to capture the intuitive requirements.

The results of Pultr previously discussed apply to *ordinary* fuzzy set theory; Katz' results develop *standard* Łukasiewicz logic (even if interpreted à la Scott). The comments of this Section, then, do not refer to extensions of the basic theories. Our conclusions on the fact that one aspect of vagueness is captured by these basic theories while the other, more innovative, ones are not, come close to what Heinz Skala (1982) writes on interactive and non interactive vagueness and to the remarks made by Joe Goguen (1974) on the plausibility of an "atomic hypothesis" for concepts. These different aspects and points of view would deserve being further analyzed and related but this cannot be done here. We shall limit ourselves to making two remarks.

First. Interaction is related to change and it is just when changes occur that vagueness manifests those aspects which are the most elusive and the most innovative at the same time. So, in order to obtain a satisfactory formal representation, vagueness calls for an adequate setting: a foundational framework able to accept a representation of the dynamics of the descriptions and of the change in the points of view, therefore represents also *conceptual variations of information* and not only *lack of information*.

Second. The innovative aspects of vagueness are probably related either to the non-existence of the final atomic constituents of concepts or, at least, to the overwhelming difficulty of picking them out. The problem of the existence of these atomic constituents of concepts must undoubtedly be eventually related to questions of meaning and we cannot but agree with Goguen that *" "meaning" has its "root" in something far deeper"* than present theories enable us to reach.
Let us conclude this subsection with some comments relating to a positive aspect of the results discussed in this Section, namely the convergence of the different trends which have been considered and which show a network of interesting connections among the different aspects of the field.

The overall picture emerging from this recognition presents a unity and a compactness which a few years ago would have been unthinkable. This looks very promising both for further theoretical developments and for applications in all those cases in which only those aspects of vagueness that are adequately captured by the tools discussed in the preceding pages are involved. Let us here summarize the main relationships. Pultr's results, as was already discussed, provide an interface between fuzzy sets and classical representations.
These latter, being of a metrical type, can be fruitfully related to the metrical developments of many-valued logic obtained by Katz; and, finally, Giles' results provide a very useful interface between set-theoretical and logical points of view obtained by means of an interpretation of dialogical type which connects theories of fuzzy sets and many-valued logics.

2.3. Parikh's *approach to vagueness*.

In this subsection we wish to briefly discuss the analysis of vague predicates made by Rohit Parikh (1977) and his formal proposal to treating them. In order to do this we must summarize some of his arguments. The starting point of Parikh's discussion is neither the presence of borderline cases nor the existence of dispersive propositions, but two "sorites"-like

paradoxes - one regarding the use of colour terms, the other the eye's power to discriminate colours - which were considered by Dummett (1975) and Wright (1975). By analyzing the paradoxes one realizes that the controversial points are related to the presence of observational predicates. More specifically, according to Dummett and Wright, the trouble resides in the fact that those words which refer to observational predicates denote vague predicates and not predicates in the sense of Frege. Parikh rightly observes that this kind of problem does not impinge only on the use of ordinary language but *"also affects physics and, in fact, any science that attempts to correlate observation with words or numbers" (page 3)*. Parikh then proposes to pursue the analysis in a suitable mathematical setting. Such a framework is provided by the α-connected and connected metric spaces. A metric space is α-connected if it cannot be divided into two disjoint nonempty subsets A and B such that the distance $d(x,y)$ between any two elements, $x \in A$, $y \in B$, of the two subsets is greater than α. A connected metric space in the usual sense obviously satisfies the previous condition for any α. The following remarks by Parikh are central to the present discussion. Let us consider a metric space consisting of (some aspects of) physical objects. The space will be called *"observationally connected if it is α-connected for some α which is smaller than the limits of observation" (page 10).*
It follows from the definition of α-connectedness that if a predicate defined on an α-connected metric space is nontrivial then there must exist at least two points whose distance is less then α and such that the predicate applies to one of these points but not to the other.

We must then face the following consequences: In an observationally connected space it is not possible to define nontrivial observational predicates. A fortiori, this is true in any connected metric space. The conclusion to which we are led by Parikh is that *"if we wish to observe and talk about such spaces, we must necessarily resort to vague predicates. And this restriction applies not only to ordinary people using their eyes and ears, but also to scientists using their instruments"*. It then appears that *"the choice is between no knowledge and vague knowledge, whatever the latter might be" (page 11).*

Vague predicates, then, seem to be essential if we wish to take into account what scientists actually do; this, however, forces us to consider the well known fact that classical logic cannot be straightforwardly applied to vague predicates. A solution to this problem must then be sought and, hopefully, a solution fitting in naturally with actual empirical activities. Parikh, moreover, observes that taking into account the problems of vagueness, approximation and imprecision in empirical measurements forces us to reconsider the received view on the relationship existing between mathematical theories and the "world of actual measurements". The correspondence between mathematical theories and the empirical world (undoubtedly existing at a certain level, otherwise the effectiveness of scientific theories would remain unexplicable), cannot be pushed too far since it is also undeniable that some questions which are of central interest in the mathematical theories seem to have no relevance in scientific disciplines and viceversa. Therefore a conception which considers the activity of experimental scientists when involved in the problems of measurements as an approximation of what a corresponding mathematical theory states, is untenable. This perspective is very interesting since it stresses a similarity existing among naked eye observations, operations of measurement of everyday life, scientific empirical measurements and also the use of ordinary language. Parikh then presents a Theory of Vague Reals (TVR) aimed at avoiding the inconveniences to which vague predicates give place. TVR accounts for the process of actual measurements and its logical structure has an intuitionistic flavour; it is possible to define vague predicates in terms of TVR but "Sorites"-like paradoxical situations cannot occur since no predicate has a classical negation.

All the problems seem, then, solved; unfortunately there is one limitation. TVR can be used only in those situations *"where measurements can, at least in theory, be arbitrarily accurate" (page 29)* since it makes use of rational intervals which can be arbitrarily small. It is, then, not possible to apply TVR to cases in which there is *"an inherent limit to accuracy"*. But this is the common situation of vague predicates defined in terms of direct observation (does something similar apply also to some quantum predicates?). Thus we have solved the problem of empirical measurements but not the problem of vague predicates in general. The reason why vague predicates considered at a phenomenal level cannot be accounted for easily in a theory of intuitionistic type, as TVR is, resides, as Parikh observes, in the following fact.

Let us suppose that some questions can be answered at a certain level of accuracy at a certain time; if the accuracy can be, in principle, improved, then it makes sense to leave some questions unsettled. An answer which is neither clearly positive nor clearly negative at a certain level of accuracy can be, in fact, definitely answered by improving the accuracy. When however, vague predicates are considered at a phenomenal level there is no future improvement to which one can refer and then it makes no sense to reserve judgement. We must classify as negative or positive also those answers which are not definitely clear, since there is no hope that conditions will change, providing additional information to allow us to decide without doubts. Parikh concludes then that *"in trying to decide if objects are red by looking at them, we are undertaking a task which is rather similar to that of constructing a recursive extension of formal number theory. Any such extension must be inconsistent and so must our use of colour terms"*.

One cannot but agree with Parikh's conclusions. These, however, leave us again in a mood similar to that of the preceding subsections. Everything is perfectly in order and the technical results obtained seem to fulfil the major part of the initial requests satisfactorily. However it seems that we have nevertheless missed capturing some central features at the core of vagueness. This feeling is reinforced if we consider the classical result of David Miller to which we referred in Subsection 2.1. This result regards specifically another notion having a "family resemblance" with vagueness; the notion of "accuracy". We refer to the paper by Miller reprinted in the present volume for a more complete discussion of the problem. Let us summarize this result here by saying that "except in the case in which only one parameter is taken into account - the degree of accuracy of a scientific theory turns out to be not invariant under trivial mathematical reformulations of the considered parameters. A certain hypothesis, then, which is uniformly more accurate than another in a certain formulation becomes uniformly less accurate than the other in another formulation". The previous result undermines the very basis of any attempt at constructing a theory of approximation, if this theory has to take into account - as it should - also the problem of numerical accuracy. (Indirectly this result also bears on any theory of vagueness if the theories of these notions - as would be desirable - must be related, as the corresponding notions are related conceptually). Even if, in fact, as Miller writes, there is nothing in the result which *"proves....that there cannot be degrees of error, or degrees of inexactness, or (looking at it from the other end) degrees of truth"*, however this result shows that we are far from solving this problem and also from planning or envisaging research strategies for its solution since if these degrees of error exist *"they depend on factors of which we have at present little cognizance"*.

A way of neutralizing the negative impact of Miller's theorem would be by providing independent reasons for choosing some parameters as more meaningful than others in order to *"make sense... of the degree to which some hypothesis, in answer to some problem, approach what might be thought to be the true answer to that problem"*. Now, of course, there are in principle, no reasons why this cannot be done. However, this position is tenable only inside a conceptual setting and a foundational framework which are based on (and show the fruitfullness of taking) an attitude which stresses the relativity of various descriptions but provides at the same time the tools for passing, in an objective way, from one description to another. This consideration allows us to return to Parikh's conclusions to see whether they can be used to provide some suggestions for further developments in a setting like the one previously outlined.

As we have already seen, a central condition for the applicability of Parikh's TVR is the possibility of having, in principle, unlimited accuracy; this, in fact, guarantees the possibility of reserving judgement in all the cases in which a question is posed when the answer can be obtained only with limited accuracy. It would be desirable to extend the possibility of reserving judgement also to the case in which predicates related to direct observations or ordinary language situations are involved; however, as already argued, this cannot be based on future improvements of accuracy. The role played by this last notion could however be fruitfully played – in a setting similar to the one outlined when discussing Miller's theorem – by the "change of point of view" (relating this to Vopěnka's (1979) observation that *"our capacity for observation and distinction is limited by the horizon in all directions" (page 39)*). One could reserve judgment also when considering for instance predicates related to direct observations since a certain point of view is too general or is not the applicable, right one to settle the question definitely.

The perspective of developing theories in which the point of view can be specifically taken into account may help in giving more firm foundations to the observation that ordinary language definitely works even if it seems to be eventually inconsistent. When using ordinary language, in fact, one has usually in mind that any discourse presupposes a class of implicit assumptions shared by the speakers. We must admit that these comments are very vague indeed. Their only aim is to stress, once again, the fact that any analysis of vagueness seems to point to the need for looking anew at some basic foundational problems. Any technical proposal, however ingenious, is unlikely to solve all the problems posed by vagueness.

3. *Concluding Remarks*

The discussion of the previous Section seems to favour the conclusion that simple solutions to the problems or paradoxes related to the presence of vague predicates are unlikely to survive the scrutiny of a detailed analysis, which wishes to preserve their innovative character.

Dummett's (1975) remark is clarifying: *"We feel that certain concepts are ineradicably vague... to take vagueness seriously is to suppose that a vague expression may have a completely specified, albeit vague, sense" (page 258).* One of the positive features of vague predicates is their ability to be extended in domain in which they neither definitely apply nor definitely not apply; their presence, then, is likely to create discrepancies between what is represented and the language used, (the representation), discrepancies which, when amplified, can produce also paradoxes.

Many people have insisted that the presence of unregimented vagueness must eventually produce a fundamental incoherence and that this incoherence is, in fact, a feature of ordinary language. Let us look at this problem from a particular point of view. It is a common experience of any reader of literary or poetic texts that an essential use of vagueness is made in order to produce an incoherence of the language which is intrinsic at the surface, literal level (and, to this aim, sometimes vagueness is purposely produced).

This fact, however, is of no relevance here since we are not considering the language of poetry but the language of science. Our claim is that there is a use of vagueness in the language of science that is essentially similar, even if in the recipe used by the scientists there is also another ingredient.

Scientists are well aware of the fact that notions and concepts which have a very precise application in a given domain and inside a determined framework become vague outside their proper domain and framework and they, in fact, usually take great care in controlling domain and framework of application. There is, however, a case in which the unorthodox procedure of using some concepts outside their original domain is intentionally used. This happens in the initial stages of a new theory during which the normal practice of the scientists is what J.L. Synge (1965) has characterized as the "cuckoo process": *"The eggs are laid, not on the bare ground to be hatched in the clear light of Greek logic, but in the nest of another bird, where they are warmed by the body of a foster mother...this cuckoo process follows the true order of historical development in science and has the advantage that at every stage of the transformation the learner has the comforting support of familiar surroundings... but it is confusing".* Referring specifically to Newtonian mechanics and Relativity, Synge continues *"the concepts of Newtonian physics interlock with one another and until one has finally reviewed all Newtonian concepts, there is always present a suspicion that the same word is being used with meanings which differ with the context". (pp. 2-3).*

Thus the embryo of the new theory is inserted into the body of a previous theory borrowing freely from the latter whichever notion can be useful. Of course, this process is likely to create discrepancies (the "confusion" of which Synge writes); where this occurs the scientist tries to locate, and then eliminate, the vagueness produced. This last point is therefore what makes a real difference between the use of vagueness in poetry and in science. In both, the possibility which vagueness offers of shifting and bending concepts is used. However in poetry one wishes these features to remain, stimulating analogical associations and allowing different levels of reading and, then, open-ended interpretations; in science, on the contrary, this flexibility of vague notions is used only as an informal heuristic tool to arrive at the construction of a new theory. Let us, incidentally, observe that these points seem to come closer to two topics of the contemporary philosophy of science, namely *"the commensurability of theories"* and *"the dynamics of science"*. We wonder whether these could be somewhat clarified if looked at from the previous point of view.

Let us return to the problem of the essential incoherence of ordinary language with respect to the problem of vagueness. Dummett (1975) writes: *"A satisfactory account of vagueness ought to explain two contrary feelings we have: that expressed by Frege that the presence of vague expressions in a language invests it with an intrinsic incoherence; and the opposite point of view contended for by Wittgenstein, that vagueness is an essential feature of language" (page 258).*

The behaviour of the scientists comes surprisingly close to both of the antithetic attitudes described by Dummett. Their leading principle in using the cuckoo process is that "vagueness must be fruitfully used but finally rejected". In doing so they are aware (with Wittgenstein) that "vagueness is an essential feature of language" and use this feature to extend previous theories; but they are also aware (with Frege) that "vague expressions in a language invest it with an intrinsic incoherence" and, in fact, they try to eliminate it.

Thus what the notion of vagueness points to, if taken seriously, is that considering non equivalent points of view simultaneously can be useful and perhaps necessary in many situations. Let us remember, as more or less unrelated examples, (a) the contrapposition between arithmetic and geometry as outlined by Bernays[6]; (b) complementarity in physics; (c) the more or less open contrapposition in Theoretical Computer Science between cathegory-theoretical and combinatorial foundations.

What vagueness is looking for is a foundational and formal framework in which it is possible to account for different descriptions. But, in fact, by accepting Russell's position that vagueness is born of a discrepancy, it is not unnatural to insist on the fact that a central and essential feature of vague predicates is the necessity of taking constanty into account two, non equivalent points of view, borderline cases being essentially produced by the clashing of the "natural" representations produced by these points of view.
All these ideas seem to fit in very naturally with some general conceptual assumptions of the Alternative Set Theory

(AST) (see Vopěnka (1979) and Sochor (1983)); in fact we had this theory in mind when developing many points of this paper. We defer to a future occasion, however, a specific analysis of the relationships between the assumptions of AST and those required by the notion of vagueness.

We can then conclude the present paper with the following remarks:
(a) If only the static aspect of vagueness is taken into account, then the results of metrical type commented in Section 2, seem to indicate a completely satisfactory direction of development.
(b) If the dynamical character of vagueness is considered, then a revision of the foundations seems to be needed, some form of constructivism seeming to be a requirement which one cannot avoid.

Istituto di Cibernetica del CNR
80072 Arco Felice, Napoli, Italy.

NOTES

The "motto" is taken from M. Dummett, Frege, Philosophy of Language, Duckworth (1981), page 647.

1. Parrat's remarks are undoubtedly related to the conviction that the uncertainty present in the scientific activities can be satisfactorily accounted for through the theory of probability. I think, nevertheless that his considerations are relevant also in those cases in which the uncertainty cannot be attributed to statistical variations.

2. Aristotle's passage to which I refer is from Nicomachean Ethics (1094b). I am indebted to Bertil Rolf for this quotation. Popper's passage which is relevant here is from his Autobiography contained in vol. XIV, Book 1 of the Library of Living Philosopher edited by P.A. Schilpp, page 17: "I do not suggest, of course, that an increase in the precision of, say, a prediction, or even of a formulation, may not sometimes be highly desirable. What I do suggest is that it is always undesirable to make an effort to increase precision for its own sake - especially linguistic precision - since

this usually leads to loss of clarity...: one should never try to be more precise than the problem situation demands.... an increase in precision or exactness has only a pragmatic value as a means to some definite end".

3. One must explicitly say that, in fact, they have dealt at no length with the problem of vagueness but the two brief passages referred to in Note 2 above are explicit enough to allow us to speak of their attitude towards the problem of precision.

4. See Dummett (1975), page 260 of the 1978 reprint.

5. See Russell (1923) page 85. Let us remark that Russell wrote on this problem only few pages which, however - as well as the few comments of Aristotle and Popper - are of lasting value.

6. See his 'On Platonism in Mathematics', l'Enseignement Mathematique, 34 (1935), pp. 52-69 (in french), translated into english in P. Benacerraf and H. Putnam (eds.), Philosophy of Mathematics, Basil Blackwell, Oxford, 1964.

Acknowledgments

Mike Briggs-Smith greatly helped in reducing the linguistic clumsiness of the text. I am very grateful to him for his kindness and patience. Thanks are due to A.M.Mazzarella for her skillful typing of the paper.

REFERENCES

Dummett, M.: 1975, 'Wang's paradox', Synthese 30, 301-324, reprinted as chapter 15 of his Truth and Other Enigmas Harvard University Press, Cambridge, Mass. 1978.

Gaines, B.R.: 1976, 'Foundations of Fuzzy Reasoning' International Journal of Man-Machine Studies, 8, 623-668.

Giles, R.: 1976, 'Łukasiewicz Logic and Fuzzy Set Theory' International Journal of Man-Machine Studies 8, 313-327.

Goguen, J.A.: 1974, 'Concept Representation in Natural and Artificial Languages', International Journal of Man-Machine Studies 6, 513-561.

Gottwald, S.: 1983, 'Fuzzy Set Theory, Some Aspects of the Early Development' (this volume)

Katz, M.: 1981, 'Two Systems of Multi-valued Logic for Science' Proc. XI Symposium on Multiple-Valued Logics, Oklahoma City, 175-182.

Katz, M.: 1982, 'Real-valued Models with Metric Equality and Uniformly Continuous Predicates', Journal of Symbolic Logic 47, 772-792.

Ludwig, G.: 1981, 'Imprecision in Physics' in A. Hartkämper and H. J. Schmidt (eds.), Structure and Approximation in Physical Theories, Plenum Press, New York pp.7-19.

Miller, D.: 1983,'Impartial Truth' (this volume)

Parikh, R.: 1977, 'The Problem of Vague Predicates' (preprint) Boston University.

Parrat, L.G.: 1961, Probability and Experimental Errors in Science, J. Wiley and Sons, New York.

Pavelka, J.: 1979, 'On Fuzzy Logic', Zeitschrift für Math. Logik und Grundlagen der Math 25, 45-52, 119-134, 447-464.

Pultr, A.: 1983, 'Fuzziness and Fuzzy Equality' (this volume)

Russell, B.: 1923, 'Vagueness', Australian Journal of Philosophy, 1, 84-92.

Scott, D.: 1976, 'Does Many-valued Logic has any use?' in S. Körner (ed.) Philosophy of Logic, Basil Blackwell Oxford pp. 64-74.

Skala, H.J.: 1982, 'Modelling Vagueness' in M.M. Gupta and E. Sanchez, Fuzzy Information and Decision Processes North-Holland Publ. Co., Amsterdam pp. 101-109.

Sochor, A.: 1983, 'The Alternative Set Theory and Its Approach to Cantor's Set Theory' (this volume)

Synge, J.L.: 1965, _Relativity, the Special theory_, North-Holland Publ. Co., Amsterdam.

von Neumann, J.: 1956, 'Probabilistic logics and the Synthesis of Reliable Organisms from Unreliable Components' in C.E. Shannon and J. McCarthy (eds.), _Automata Studies_, Princeton University Press pp. 43-98.

Vopěnka, P.: 1979, _Mathematics in the Alternative Set Theory_ Teubner Verlagsgesellschaft, Leipzig.

Wright, C.: 1975, 'On the coherence of vague predicates', _Synthese_ 30, 325-365.

Zadeh, L.A.: 1975. 'The Concept of a Linguistic variable and its Application to Approximate Reasoning', _Information Sciences_ 8, 199-249, 301-357; 9, 43-80.

E. Trillas and L. Valverde

AN INQUIRY INTO INDISTINGUISHABILITY OPERATORS

1. INTRODUCTION

Discerning objects in a theory is an essential problem. Quine (1981) has said "There is no entity without identity". The problem of individuation is a long-standing philosophical preoccupation (Munitz,1971); because though things are, themselves,what they are, a human or artificial observer aided by intelligence or instruments "reads" things and sees in them just what can be seen, what the discriminating power of the instrument allows, this being true for physical as well as intellectual instruments.

We make here a proposal that sheds a common light on problems arising in various fields and lets them share a common mathematical structure. The proposal is presented in mathematical language and originates in assuming a previous "equality" whereupon the concept of "indistinguishability" is built. The later notion becomes necessary when the language we use is richer because we want it to cover vagueness. Such a language also forces a new "classification" of objects which does not always coincide with the idea of a (sharp-edged) partition. The biconditional in Classical Logic, equivalence in Three-valued Logic, indifference as double preference in the Social Sciences, indistinguishability in Probabilistic Metric Spaces, Menger's probabilistic relations of indistinguishability, Zadeh's similarity relations, Ruspini's likeness relations for Clustering, and others, are problems requiring an additional distinction.

As long as present learning theory becomes increasingly a scientific subject, technical interaction among people from various fields is becoming more and more important to find out which are the laws governing the continuing process of mental ordering and classification of objects. There is a growing suspicion that the human mind confronts the problem of "Knowing" spurred by the empirical shock of the senses and the phisical world. The mind seems to proceed by delineating knowledge at increasingly finer levels through orderings and classifications, which differ from those made in ordinary life and prompt

H. J. Skala, S. Termini, and E. Trillas (eds.), Aspects of Vagueness, 231–256.

a "logic" required to follow closely the contours of the knowledge sought. Reichenbach (1947) says "once a result of thinking is obtained, we can reorder our thoughts in a cogent way, constructing a chain of thoughts between points of departure and point of arrival; it is this rational reconstruction of thinking that is controlled by logic, and whose analysis reveals those rules which we can call logical laws". We do such orderings and classifications as best as we can. In the long run, they bear out a type of logic which can be deduced from the whole process.

For mental acquisition and systematization of knowledge in natural or artificial intelligence the required effort is basically interdisciplinary and of the "applied logic" type, in Enriques' sense (see Ullmo,1969). "Logic" surely because of the "implication" character of many indistinguishability operators. "Applied" because, though both methodology and language used are clearly mathematic, it is only through interaction and preoccupation with "things" that the mathematical models are obtained.

1.1 Field cases suggesting the indistinguishability operator concept.

We now proceed to survey some situations, of differing degree of familiarity, where an indistinguishability operator arises naturally. The indistinguishability the operator is intended to model overlays identity -it does not supplant it- and the modeling is adjusted by considerations on the behavior -or some additional meaning- of the objects involved. The focus of the survey is on the case study of an operator, not on the case analysis itself.

1.1.1. Indistinguishable random events.

If X is an universe of discourse and $\mathcal{R}$ is a Boolean algebra of parts of X modeling a random experience of samples in X, then the "events" in $\mathcal{R}$ can be more relevantly analyzed as indistinguishable -in the random sense- than as set-theoretical "equal" subsets of X.

Two events A and B are "indistinguishable random events" if <u>one almost never coincides with the complement of the other</u>. So, if p is a properly defined probability in $\mathcal{R}$, p-indistinguishability between A and B can be defined by $p(\overline{A}\cap B)$ $=p(A\cap\overline{B})=0$. This amounts $p(A\Delta B)=0$. Thus, the $[0,1]$-valued operator $E_p(a,b)=1-p(A\Delta B)$, defined in $\mathcal{R}\times\mathcal{R}$ <u>is a degree of in-</u>

distinguishability of two random events and the maximal value $E_p(A,B)=1$ characterizes two events as random-indistinguishable.

Since "$A=B$ iff $\overline{A \Delta B}=X$" it seems clear that random indistinguishability can be superimposed on object identity by the added language of the given random experience (e.g. dice throwing). Events are what they are, yet random experience compellingly introduces a new concept that is out of the reach of set theory in which they are mathematically modelled.

Notice that if A and B are p-independent, then $E_p(A,B) =1-p(A)-p(B)+2p(A)p(B)$. It should be recalled that $p(A\Delta B)$ defines a well-known event distance in $\mathcal{R}$. (Neveu, 1964).

1.1.2. Congruent geometric figures.

If X is a "set of points" and $\mathcal{F}=\{F,G, \ldots\}$ is a distinguished family of parts of X we call figures, then to make geometry the "identity" between figures: "$F=G$ iff $\mu_F(x)=\mu_G(x)$ for any x in X" (where μ_F represents the characteristic function of F) does not suffice. The new geometric concept of "congruence" is required. We thus introduce a set Σ of transformations $\sigma:X \to X$ (we better choose it to be a group) and we define "F and G are congruent modulo Σ" and write $F\equiv G\ (\Sigma)$ if there exists some σ in Σ such that $\sigma(F)=G$. This definition entails the use of the two quantifiers; this is clearly seen when using the characteristic functions, so we have

$$\text{"}F\equiv G\ (\Sigma) \text{ iff } (\exists\sigma\in\Sigma)(\forall x\in X: \mu_{\sigma(F)}(x)=\mu_G(x))\text{"}.$$

With the Kronecker δ, the operator

$$E_\Sigma(F,G)=\sup_{\sigma\in\Sigma}\ \inf_{x\in X}\ \delta(\mu_{\sigma(F)},\mu_G(x))\in\{0,1\},$$

let us write "$F\equiv G\ (\Sigma)$ iff $E_\Sigma(F,G)=1$". If the Σ group contains only the identity, "$F\equiv G\ (\Sigma)$ iff $F=G$" is obtained, so we remark that figure identity is a special case of congruence. Since $\delta=1-d_o$, where d_o is the discrete distance in X, we have

$$E_\Sigma(F,G)=1-\sup_{\sigma\in\Sigma}\ \inf_{x\in X}\ d_o(\mu_{\sigma(F)}(x),\mu_G(x)),$$

where the indistinguishability operator is represented as a function of a distance.

1.1.3. Indistinguishability of "nebulae" in a screen.

Let X be a screen where nebular clusters $\underset{\sim}{F},\underset{\sim}{G}$, ... of points appear (e.g. stains made by perpendicular spray-painting from a regular distance). We call these clusters "nebulae". We will suppose they classify the points of X into three classes: those in the nebula, those not in the nebula and those with the indeterminate status of being in or out of the nebula.

Each nebula can thus be characterized by a "generalized characteristic function" $\mu_{\underset{\sim}{F}}$ from X into $\{0,1/2,1\}$, so that $\mu_{F}(x)=1$ if the point x is clearly in $\underset{\sim}{F}$, $\mu_{\underset{\sim}{F}}(x)=0$ if it is not clearly in $\underset{\sim}{F}$ and $\mu_{\underset{\sim}{F}}(x)=1/2$ otherwise. We can then extend the Kronecker δ by defining $\delta(0,1/2)=\delta(1/2,0)=\delta(1,1/2)=\delta(1/2,1)=1/2$, $\delta(1/2,1/2)=1$, which covers the indeterminate cases. We have $\delta(x,y)=1-|x-y|$. By analogy with case 1.1.2 we can consider the operator

$$E(\underset{\sim}{F},\underset{\sim}{G})=\inf_{x\in X}\delta(\mu_{\underset{\sim}{F}}(x),\mu_{\underset{\sim}{G}}(x))=1-\sup_{x\in X}|\mu_{\underset{\sim}{F}}(x)-\mu_{\underset{\sim}{G}}(x)|\in\{0,1/2,1\},$$

that gives the degree of relative indistinguishability between two nebulae $\underset{\sim}{F}$ and $\underset{\sim}{G}$. Then:

a) $\underset{\sim}{F}$ and $\underset{\sim}{G}$ are distinguishable if $E(\underset{\sim}{F},\underset{\sim}{G})=0$. That means that $\sup_{x\in X}|\mu_{\underset{\sim}{F}}(x)-\mu_{\underset{\sim}{G}}(x)|=1$, and therefore that <u>there exists some point x in X which is in a nebula and not in the other</u>.

b) $\underset{\sim}{F}$ and $\underset{\sim}{G}$ are indistinguishable if $E(\underset{\sim}{F},\underset{\sim}{G})=1$. That means that $\sup_{x\in X}|\mu_{\underset{\sim}{F}}(x)-\mu_{\underset{\sim}{G}}(x)|=0$, and therefore that for any point in X $\mu_{\underset{\sim}{F}}(x)=\mu_{\underset{\sim}{G}}(x)$: <u>both nebulae $\underset{\sim}{F}$ and $\underset{\sim}{G}$ coincide</u>.

c) $\underset{\sim}{F}$ and $\underset{\sim}{G}$ are <u>mixed up</u> if $E(\underset{\sim}{F},\underset{\sim}{G})=1/2$. That means that $\sup_{x\in X}|\mu_{\underset{\sim}{F}}(x)-\mu_{\underset{\sim}{G}}(x)|=1/2$, and therefore that <u>there is no point that is in a nebula and not in the other and there exists at least one point that doubtfully belongs to a nebula but that undoubtedly belongs to the other</u>.

Notice that the $|x-y|$ distance ought to be modified if, for instance, the screen had deformations in some regions

1.1.4. Statistically indistinguishable probability laws.

A <u>statistical structure</u> (Barra,1971) on a measurable space $(X,\mathcal{R})$ is given by a family $\{p_\theta,\theta\in\Theta\}$ of probabilities defined on the $\mathcal{R}$ algebra. Statistical problems are characterized by the hypothesis that a probability is involved but it is unknown which (out of the p_θ's). Given the statistical structure $(X,\mathcal{R},\{p_\theta,\theta\in\Theta\})$ and an event $A\in\mathcal{R}$, to say that two probabilities are both equally useful amounts to saying that there are statistically indistinguishable probabilities.

Given A, each time we have a normalized distance d_A defined on $\{p_\theta, \theta \in \Theta\}$ we can give a degree of indistinguishability relative to A through $E_A(p_\theta, p_{\theta'}) = 1 - \sup d_A(p_\theta, p_{\theta'}) \in [0,1]$. By analogy with the former cases, the degree of of indistinguishability can be then defined by

$$E(p_\theta, p_{\theta'}) = \inf_{\mathcal{R}} E_A(p_\theta, p_{\theta'}) = 1 - \sup_{\mathcal{R}} d_A(p_\theta, p_{\theta'}) \in [0,1].$$

Thus, in the special case we had $d_A(p_\theta, p_{\theta'}) = |p_\theta(A) - p_{\theta'}(A)|$, we would have $E(p_\theta, p_{\theta'}) = 1 - \sup_{\mathcal{R}} |p_\theta(A) - p_{\theta'}(A)|$, then $E(p_\theta, p_{\theta'}) = 1$ is equivalent to the fact that $\sup_{\mathcal{R}} |p_\theta(A) - p_{\theta'}(A)| = 0$, so $p_\theta(A) = p_{\theta'}(A)$ for any event A in $\mathcal{R}$, i.e. both laws coincide.

Once again we find that the objects involved (here the p_θ probabilities) are perfectly defined and that the indistinguishability appears when a second language (here the language of statistics) is required to model mathematically the actual situation.

1.1.5. Approximate synonymy.

Given a universe of discourse X and a set of linguistic terms, if we call "relational language" a binary relation L between X and T and we define à la Zadeh (Zadeh,1971) the extensional meaning of a term t in T as the subset $\sigma_t = \{x \in X; (x,t) \in L\}$ of X, then the classical definition of synonymous terms (over the linguistic structure (X,T,L) is "t is a synonym of t' ($t \sim t'$) iff $\sigma_t = \sigma_{t'}$" . In order to combine linguistic terms logically, and significantly, we set forth the hypothesis that the set $\mathcal{R} = \{\sigma_t; t \in T\}$ has some proper set-theoretical structure with the basic operations of union, intersection and complement.

Given a subset $S \subset X$ that corresponds to no term in T (i.e. no term t in T exists such that $\sigma_t = S$), if for instance t and t' exist such that $\sigma_t \subset S \subset \sigma_{t'}$ and $S - \sigma_t$, $\sigma_{t'} - S$ are "comparatively small" then S can be said to have "aproximate terms" t and t' that turn out to be mutually "approximate synonyms" and that are useful to mean approximately S. We will write $t \approx t'$. With approximate synonymy we have the standard linguistic phenomenon of the breaking of synonymic sequences; i. e. given a sequence $t_1 \approx t_2$, $t_2 \approx t_3$, ..., $t_{n-1} \approx t_n$, we can yet have $t_1 \not\approx t_n$, contrasting with the case of "exact" synonymy from the extensional definition above. Actually, this vagueness phenomenon -aparently inevitable in natural languages- is motivated by the need we have of naming objects with a limited vocabulary.

What happens is that in natural or near-natural languages synonymy is a matter of degree rather than an absolute quality. If $\mathcal{B}$ is a Boolean algebra and a probability is avalaible, we can define the index of synonymy between terms <u>(linguistic indistinguishability)</u> as the number $\rho(t,t')=1-p(\overline{\sigma_t \Delta \sigma_{t'}})$, for which we have " $\rho(t,t')=1$ iff $p(\overline{\sigma}_t \cap \sigma_{t'})=p(\sigma_t \cap \overline{\sigma}_{t'})=0$", so that if $(\mathcal{B},p)$ has no negligible event other than ϕ, the classical case $\sigma_t=\sigma_{t'}$ is obtained.

If no significant probability is available, or if $\mathcal{B}$ is not a Boolean algebra, and if X is finite, then we can define the <u>index of synonymy</u> as the number $\rho^*(t,t')=|\sigma_t \cap \sigma_{t'}|/|\sigma_t \cup \sigma_{t'}|$ ($|\sigma_t|$ is the number of elements of σ_t), satisfying:
a) $\rho^*(t,t')=\rho^*(t',t)$; b) $\rho^*(t,t')=1$ iff $\sigma_t=\sigma_{t'}$ iff $t \approx t'$, and c) $\rho^*(t,t')=0$ iff $\sigma_t \cap \sigma_{t'}=\phi$.

Such a definition suggests an interesting extension: if we assume, in order to catch the vagueness which is inherent in approximate synonymy, that language L is a fuzzy relation, $L: X \times T \to [0,1]$, then the "extensional meanings" will be possibility distributions given by $\sigma_t(x)=L(x,t) \in [0,1]$, to be combined logically according to agreed-upon rules. In particular, ρ^* will have an expression that will depend on the functional expression of the two logical connectives $\cap$ and $\cup$. Should we choose Łukasiewicz's $\aleph_1$ logic, we would have:

$$\rho^*(t,t')=\frac{\sum_{i=1}^{n} \max\{0,\sigma_t(x_i)+\sigma_{t'}(x_i)-1\}}{\sum_{i=1}^{n} \min\{1,\sigma_t(x_i)+\sigma_{t'}(x_i)\}} ,$$

(where $X=\{x_1,x_2, \ldots, x_n\}$). In that case, an easy calculation leads to " $\rho^*(t,t')=1$ iff $(1/n)\sum_i|\overline{\sigma}_t(x_i)-\sigma_{t'}(x_i)|=1$", i. e. <u>the index of linguistic indistinguishability between terms with vage meaning is a maximum if, and only if, the Hamming</u>

distance between the meaning of the first and the complement of the meaning of the other is a maximum, in the context of Łukasiewicz $\aleph_1$ logic.

Again the need we have of using an additional, not strictly formalized, language forces us to view the formal identity of objects pragmatically as an indistinguishability operator (that turns out to be strongly related to a distance).

1.2. A comment on vague predicates and indistinguishability operators

Though the purpose of the present article is not the logical analysis of the so-called vague predicates (however basic they are for indistinguishability relations) it seems appropiate to finish the given list of examples with a guided tour to their mathematical aspects suggested by Parikh's paper (Parikh, 1983).

To say that X is a "set" presupposes that certain objects x,y,z, ... exist and that we have some method to decide, every time we consider one of them, whether it does or does not belong to X. If & is any such object, the "& belongs to X" is a Fregean predicate and the query "does & belong to X?" admits only one answer, either yes or no.

Objects replacing & must be mutually distinct, and mathematicians are very strict about this. A great deal of Mathematics is consecrated to the formal making of its own building blocks. The mathematicians' imagination has worked hard not to interfere with the generally unsolvable problem of writing (naming) all the elements in the classes they consider.

One of the most useful instances is the real number. If we consider π, a tipically geometric number (because reflecting a metric property of circumference), we can define it by the following Theorem: For any circumference C, if l is its length and d the length of any of its diameters, then the l/d quotient is a real number named π. To obtain this theorem we must use both Mathematical Analysis and Euclidian Geometry.

Mathematicians have invented (mathematical) methods to compute π in some numbering system, but since π is a trascendental real number (this being a theorem necessitating the whole Mathematical Analysis) it turns out it is not possible to write it complete. We do not seem to have other options but finally stating "π is π, and nothing more". Actually the final statement is almost invariably "π is an object in an open interval (e.g. (3.1415,3.1416)) with rational endpoints".

Thus the writing of π is an abstraction over a family of open intervals (a,b) such that $3.1<a<b<3.2$, a,b in Q, and such that we can assure we can make the differences $|\pi-a|$, $|\pi-b|$ as arbitrarily small as possible: the computing mathematician will halt the calculation as soon as

$$\max\{|\pi-a|, |\pi-b|\} \leqslant \varepsilon/2,$$

where $\varepsilon/2$ is a certain previously fixed indistinguishability level, which is determined by the ability to distinguish the interval's endpoints. Since $|a-b| \leqslant |\pi-a| + |\pi-b| \leqslant 2\max\{|\pi-a|, |\pi-b|\}$, we have $|a-b| \leqslant \varepsilon$. Thus, once the $\varepsilon/2$ level is fixed, the calculation can halt in practice as soon as an interval (a,b) is avalaible of length ε.

How do we impose the $\varepsilon/2$ level?. It depends on the "phisical ability" to distinguish pairs of points. It is generally assumed that some physical (graphical) representation is avalaible; in the case of numbers the representation is a straight line provided with a unit or measurament scale, where differences $|a-b|$ are "seen" (by the naked eye or by a suitable scale-adjusted optical apparatus as in precision scaling) as a function of this scale, with a certain indistinguishability level. In any case a control design is needed that restricts the possibilities of distinguishing between points. The point distinction can be uniform (in the case of straight line unit) or non-uniform (in the case of a curve and a magnifying glass that deform edges) and, in each case, the controlling device should reflect the fact.

Thus, starting from some formal constructs we arrive at the need of a physical model. This new model turns the mathematical non-vague predicates we had into new observational vague predicates. Therefore, we speak of new objects (a family of open intervals (a,b)) different from the π number for which the $2.\max\{|\pi-a|, |\pi-b|\}$ operator does not exceed a certain bound. If we write

$$E_\pi(a,b) = 1 - 2.\max\{|\pi-a|, |\pi-b|\},$$

since $E_\pi(a,b) \leqslant 1-|a-b|$, as soon as $1-\varepsilon \leqslant |a-b|$ we will have $E_\pi(a,b) \leqslant \varepsilon$, and a,b will be indistinguishable (in the π calculation).

The E_π operator contains the "nasty" variable π, which happens to be the cause of the problem. Luckily the problem is solved by globalizing, so it becomes independent of π by virtue of the fact that the $E(a,b) = 1-|a-b|$ operator is greater than the E_π operator.

We have thus reformulated the problem into a mathematical treatment again, so that we can go on using the tools for which there is maximum guarantee: Mathematics.

Therefore it seems that operators of type 1-d, where d is a distance, are better candidates to model mathematically the indistinguishability that adds to identity as soon as a language, theoretical or not, is superimposed upon a formal environment.

2. ON INDISTINGUISHABILITY OPERATORS

Some operators, satisfying definite properties, have been introduced by several authors as a tool for modelizing different kinds of "indistinguishability". We start by analyzing the properties of such operators to show their common features.

Thus, in (Menger,1951) the author introduced the so-called Probabilistic Relations, in order to get "a more realistic, theoretical description of equality or equivalence of the two elements a and b". Such a goal can be attained, in Menger's proposal, by assigning to each couple of objects a and b a number E(a,b) satisfying the following properties:

P1. $E(a,a)=1$;
P2. $E(a,b)=E(b,a)$, and
P3. $E(a,b).E(b,c)\leqslant E(a,c)$,

for any a,b,c of the given universe of discourse X. E(a,b) is assumed to be the probability of finding a and b indistinguishable, i.e. the relative frequency of the cases in which a and b have not been distinguished.

On the other hand, in (Zadeh,1971b) the author defines the so-called Fuzzy Equivalences or Similarity Relations. A similarity relation on a set $M=\{1,2, \ldots, n\}$ is given by a nxn matrix $E=(e_{ij})$, $e_{ij}\in[0,1]$ for any i,j; satisfying the properties:

1. $e_{ii}=1$;
2. $e_{ij}=e_{ji}$, and
3. $\sup_M(e_{ik}\wedge e_{kj})\leqslant e_{ij}$,

for any i,j,k in M. e_{ij} is assumed to be the strength of the link between the elements i and j. Notice that the map E from MxM into [0,1] given by $E(i,j)=e_{ij}$ defines an operator such that

S1. $E(i,i)=1$;
S2. $E(i,j)=E(j,i)$, and

S3. $\text{Min}(E(i,k),E(k,j)) \leqslant E(i,j)$,

for any i,j,k in M. Conversely, if E is a map from MxM into [0,1] satisfying the above properties, then the nxn matrix $(E(i,j))_{i,j}$ is a similarity relation on M. Notice that if for any i,j in M we have $E(i,j) \in \{0,1\}$, then the similarity relation given by E is nothing but a (classical) equivalence relation. In fact, any similarity relation defines a chain of equivalence relations $\{R_\lambda ; \lambda \in [0,1]\}$, being $R_\lambda = \{(i,j) \in M \times M;\ E(i,j) \geqslant \lambda\}$, for each $\lambda \in [0,1]$. It is easy to show that if $\lambda \leqslant \lambda'$ then $R_\lambda \supset R_{\lambda'}$.

We finish these introductory examples by mentioning <u>Likeness Relations</u> (Ruspini,1982). A likeness relation on a non-empty set X is a map from XxX into [0,1] satisfying the following properties:

L1. $L(a,a)=1$;
L2. $L(a,b)=L(b,a)$, and
L3. $\text{Max}(L(a,b)+L(b,c)-1) \leqslant L(a,c)$,

for any a,b,c in X. Such relations were introduced in the frameork of Fuzzy Clustering and, among other considerations, the rationale behind their definition is given by the fact that if L is a likeness relation on X, then $d(a,b)=1-L(a,b)$ is a normalized pseudo-metric on X, that is, such a relation is given (and gives) a distance between the elements of X. By means of this distance a "classification" of the objects in X is obtained.

Later we will give more examples, but now let us remark that in the above three examples the indistinguishability is given by means of operators defined on a set X with values on a ordered semigroup $([0,1],F,\leqslant)$, being $\leqslant$ the usual order of the unit interval, $F(x,y)=x.y$ in the first case, $F(x,y) =\text{Min}(x,y)$ in the second and $F(x,y)=\text{Max}(x+y-1,0)$ in the last case. Each of these operators satisfies three conditions which recall classical reflexivity, symmetry and transitivity respectively. From this standpoint, we give the following

<u>Definition 2.1</u>. Let $\mathcal{L}=(L,*,\leqslant)$ be an ordered semigroup. For a given non-empty set S and $\lambda \in L$, a map E from SxS into L is called <u>$\mathcal{L}$-indistinguishability operator on S at level λ</u> if the following properties hold for any a,b,c in S:

I1. $E(a,a) \geqslant \lambda$;
I2. $E(a,b)=E(b,a)$, and
I3. $E(a,b)*E(b,c) \leqslant E(a,c)$.

This last property, which we will call *-transitivity, should be understood in a weaker sense than the usual one, because it just expresses a threshold which should be attained by the degree of indistinguishability between a and c, given the degrees of indistinguishability between a and b and b and c. Since the operations used in the above examples satisfy $F(x,y) \leqslant Min(x,y)$ for any x,y in [0,1], it follows that a triple a,b, c of elements may exist in S such that a and b are indistinguishable at certain level, as well as b and c, but a and c are not indistinguishable at this level. Such a situation is usually present in the physical world. The associativity is imposed on these operations in order to extend the transitivity condition to more than three elements.

However, there exist important indistinguishability operators with values on semigroups others than semigroups over the unit interval. This is the case, for instance, of the biconditional in classical set theory, $A \leftrightarrow B = (A \rightarrow B) \cap (B \rightarrow A) = S - (A \triangle B)$, which can be viewed as an indistinguishability operator on P(S) at level S (the universe of discourse) with respect to the (partially) ordered semigroup $(P(S), \cap, \subset)$. Such indistinguishability operator is nothing but a particular case of the indistinguishability operator which can be defined in any Boolean ring $\mathcal{R} = (A,+,.,0,1)$, i.e. a complemented lattice $(A, \vee, \wedge, -)$ with maximum 1, minimum 0, being $a.b = a \wedge b$ and $a+b = (a \wedge \bar{b}) \vee (\bar{a} \wedge b)$. It is easy to prove that the map E from AxA into A defined by $E(a,b)=1+a+b$ is an indistinguishability operator at level 1 with respect to the semigroup $(A,.,\leqslant)$. Note that if p is an additive valuation on $\mathcal{R}$, i.e. a probability, then $E_p(a,b)=p(E(a,b))=1-p(a+b)$ is an indistinguishability operator at level 1 with respect to the semigroup $([0,1], F_Ł, \leqslant)$, being $F_Ł(x,y)=Max(x+y-1,0)$, that is E_p is a likeness relation on A.

This example provides also a way to interpret the material equivalence of classical logic as an indistinguishability operator, because, as is easy to prove, if $\mathcal{R}$ is a Boolean ring over A and f is any map from a non-empty set S into A, then $E_f(a,b)=E(f(a),f(b))$ is an indistinguishability operator on S with values on $(A,.,\leqslant)$. The material equivalence of classical logic can be viewed then as a particular case of such indistinguishability operators, where $A=\{0,1\}$, S is the set of propositions and f is the truth-valuation of S.

Last remark applies to any indistinguishability operator on a set S and any map f from a non-empty set S' into S, that is we have the

Theorem 2.1. Let E be an $\mathcal{L}$-indistinguishability operator on a set S at level λ. Given a non-empty set S' and any map f from S' into S then

$$E_f(a,b)=E(f(a),f(b)),$$

is an $\mathcal{L}$-indistinguishability operator on S' at level λ.

Thus, we can take the same standpoint of classical logic to analyze the material equivalence in Multiple-valued Logic, especially for the three-valued systems of Łukasiewicz and Bočvar. In the former, the equivalence is defined by means of $E_v(p,q)=1-|v(p)-v(q)|$, which is given by the indistinguishability operator at level 1 on $\{0,1/2,1\}$ with respect to the ordered semigroup $(\{0,1/2,1\},F_Ł,\leqslant)$ defined by $E(x,y)=1-|x-y|$ and the truth-valuation v. In the Bočvar system, we have $E_v(p,q)=1-v(p)-v(q)+2v(p)v(q)$, which is given also by an indistinguishability operator on $\{0,1/2,1\}$ with values in the same semigroup of Łukasiewicz equivalence, but now at level 1/2.

Last examples show us two essential ways in which a second "equality" has been formalized in the past: in classical logic -where there is no vagueness: objects are well-defined and known- such equality is given by an equivalence relation. Obviously, in problems involving vagueness such formalism is unsatisfactory:then the solution may be given by means of a distance, i.e. perhaps we don't know -and we can't know- if two given objects are equal according to some criteria, but we may be able to fix, according to these criteria, how far they are. In this sense the material equivalence of Łukasiewicz logic could be understood.

Now we will show in which way these two formalisms are subsumed into the indistinguishability operator formalism, that is, the following results are devoted to analyze the links between indistinguishability operators and equivalence relations as well as the links between indistinguishability operators and metric spaces. First of all, we have

Theorem 2.2. Let E be an $\mathcal{L}$-indistinguishability operator on a non-empty set S. If for any $\lambda'\leqslant\lambda$, $R_{\lambda'}=\{(x,y)\in S\times S\ ;\ E(x,y)\geqslant\lambda'\}$, then $R_{\lambda'}$ is an equivalence relation on S if, and only if, $\lambda' * \lambda'\geqslant\lambda'$.

Thus, in the case of similarity relations and biconditional in classical logic, we have that λ is the maximum of the ordered set L and, for any $\lambda'\in L$, we have $\lambda' * \lambda'=\lambda'$, consequently both indistinguishability operators give a family -indexed by L- of equivalence relations. On the other contrary, if E is either a probabilistic relation or a likeness relation, since $F(\lambda',\lambda') < \lambda'$ for any $\lambda' \neq 0,1$; only two equivalence relations are induced by the indistinguishability operator: $R_o = \{(x,y) \in SxS;\ E(x,y) \geq 0\} = SxS$, and $R_1 = \{(x,y) \in SxS;\ E(x,y)=1\}$. Notice that in case E is the indistinguishability operator giving the material equivalence of Łukasiewicz logic it is $R_1 = \Delta_S = \{(x,x) \in SxS;\ x \in S\}$.

In what concerns the link between indistinguishability operators and metric spaces we have already noticed that for any likeness relation E, $d(x,y)=1-E(x,y)$ is a normalized pseudo-distance (see Ruspini,1982). Moreover, we have:

Theorem 2.3. Let d be a pseudo-distance on a set S. Then

a) $L_d(x,y)=1/(1+d(x,y))$ is a likeness relation on S.

b) $P_d(x,y)=\exp(-d(x,y))$ is a probabilistic relation on S, and conversely.

For indistinguishability operators with respect to semigroups different from the semigroups corresponding to likeness and probabilistic relations we will not get, in general, a pseudo-metric but a generalized pseudo-metric (see Trillas and Alsina,1978). Given an ordered semigroup with neutral element e, $\mathcal{L}_e=(L,\circledcirc,\leq,e)$, a $\mathcal{L}_e$-generalized pseudo-metric on a set S is a map m from SxS into S satisfying:

GM1. $m(a,a)=e$;
GM2. $m(a,b)=m(b,a)$, and
GM3. $m(a,b) \circledcirc m(b,c) \geq m(a,c)$,

for any a,b,c in S. Then, we have:

Theorem 2.4. Let m be an $\mathcal{L}_e$-generalized pseudo-metric on a set S and let η be an order-reversing bijection from L to L. Then $E_m(x,y)=\eta(m(x,y))$ is an $\mathcal{L}'$-indistinguishability operator on S at level $\eta(e)$, being $\mathcal{L}'=(L,*,\leq)$, where $x*y = \eta(\eta^{-1}(x) \circledcirc \eta^{-1}(y))$. The converse holds only if the indistinguishability level of E is a neutral element for *.

Thus, $E(A,B)=S-(A \Delta B)$ which, as we noticed, is an indistinguishability operator on P(S) at level S with respect to the semigroup $(P(S),\cap,\subset)$ yields the generalized pseudo-me tric $m_E(A,B)=A \triangle B$, with respect to the semigroup $(P(S),\cup,\subset)$ by means of the order reversing bijection $\eta(A)=S-A$. Notice that in the above theorem we can use a different set L in order to get the semigroup in which the corresponding indistinguishability takes its values, that is, the result holds for any order-reversing bijection η from L to another given set L'. Indeed, we have used implicitily this fact in theorem 2.3: $1/(1+x)$ and $\exp(-x)$ are nothing but order reversing bijections from $\overline{R}^+=[0,+\infty]$ into $[0,1]$. We remark that the semigroup operation on $[0,1]$ obtained from $1/(1+x)$ in the way given by theorem 2.4 is not $F_Ł$ but $F(x,y)= xy/(x+y-xy)$. This fact does not contradict theorem 2.3 because we have $F_Ł(x,y)\leqslant F(x,y)$ for any x,y in $[0,1]$ and, as it is easy to prove, if E is a relation $*$-transitive, then E is $*'$-transitive for any $*'$ such that $x*'y \leqslant x*y$ for any x,y in L.

3. INDISTINGUISHABILITIES WITH RESPECT TO SEMIGROUPS OVER [0,1]

Indistinguishability operators with respect to semigroups over $[0,1]$,like similarity or likeness relations, appear in the framework of Fuzzy Set Theory as a tool for solving classification problems in which there is not a clear border between different classes, but the use of such operators can be extended in order to define an equality between fuzzy sets themselves. As it is well known, two generalized characteristic functions $\underset{\sim}{A},\underset{\sim}{B}\in \underset{\sim}{P}(X)=[0,1]^X$, i.e. fuzzy sets, are said to be equal if, and only if, we have $\underset{\sim}{A}(x)=\underset{\sim}{B}(x)$ for any x in X. Such definition implies the uniqueness of the representation of fuzzy sets by means of characteristic functions, hypothesis which is actually very strong. From this standpoint we turn now our attention to such indistinguishability operators.

We start by recalling that there exists a distinguished family of ordered semigroups over $[0,1]$ which has been extensively used in Fuzzy Set Theory: semigroups defined by continuous t-norms. As it is well known, continuous t-norms (jointly with continuous t-conorms and strong negations) has been used by several authors (see references in Valverde,1982) in order to define the meet, the join and the complementary respectively in the setting of $\underset{\sim}{P}(X)$. Thus material equivalence in $\underset{\sim}{P}(X)$ could be defined through an indistinguishability operator with respect to a semigroup given by a continuous t-norm -the meet- as material equivalence in classical and Łukasiewicz logics does. On the other hand, it is worth noting that the operations of the semigroups corresponding to either similarity relations, likeness relations or even probabilistic relations are continuous t-norms. Thus, it is not surprising that such indistinguishability operators -which we will call from now on F-indistinguishability operators- have become to the attention of several authors (see, for instance, Ovchinnikov,1982 and Valverde,1982b). Now we recall the results concerning the representation of F-indistinguishability operators in the form given by latter author. First of all we recall the following

Definition 3.1. Given a non-empty set S and a continuous t-norm F, a map R from SxS into [0,1] is called F-preorder on S at level $\lambda\in[0,1]$ if the following properties hold for any a,b,c in S:

FP1. $R(a,a)\geq\lambda$, and
FP2. $F(R(a,b),R(b,c))\leq R(a,c)$.

That is, an F-preorder at level λ on S is nothing but an F-transitive relation on S which satisfies some kind of reflexivity. Obviously, any F-indistinguishability operator is an F-preorder. Conversely, we have

Theorem 3.1. Let R be an F-preorder at level λ on S. For any continuous t-norm F' such that $F(x,y)\leq F'(x,y)$, $E(a,b) =F'(R(a,b),R(b,a))$ is an F-indistinguishability operator on S at level $F'(\lambda,\lambda)$.

Consequently, since $F(x,y)\leq Min(x,y)$ for any t-norm F, $E_M(a,b)=Min(R(a,b),R(b,a))$ is the greatest F-indistinguishability operator deduced from the F-preorder R. On the other hand, the last theorem tell us that in order to get a representation for F-indistinguishability operators it suffices to represent F-preorders. Such a representation (for indistinguishabilities at level 1) is given in the following theorem, where F is a continuous t-norm, $\hat{F}(.|.)$ its quasi-inverse, i.e. $F(x|y)=\sup\{z\in[0,1];F(x,z)\leq y\}$, and S is a non-empty set:

Theorem 3.2. Let R be a map from SxS into [0,1] such that R(a,a)=1 for any a in S. R is F-transitive if, and only if, there exists a family $\{h_j;\ j\in J\}$ of arbitrary functions from S into [0,1] such that

$$R(a,b)=\inf\{\hat{F}(h_j(a)|h_j(b));\ j\in J\},$$

for any a,b in S.

From this theorem and the above remark it follows immediately the

Theorem 3.3. Let E be a map from SxS into [0,1] such that E(a,a)=1 for any a in S. E is an F-indistinguishability operator on S if, and only if, there exists a family $\{h_j;\ j\in J\}$ of functions from S into [0,1] such that

$$E(a,b)=\inf\{F(\max(h_j(a),h_j(b))|\min(h_j(a),h_j(b)));\ j\in J\},$$

for any a,b in S.

This result also provides a representation for any generalized pseudo-metric (with respect to a continuous t-norm) on a set S, as it follows by applying theorem 2.4, taking as the order-reversing bijection from [0,1] into itself a strong negation function.

On the other hand, we remark that this theorem is a very useful tool in the fuzzy clustering processes (Ruspini, 1982). For given a sample S to be classified and a family of prototypes $\{a_1,a_2,\ldots,a_n\}\subset S$, if h_i is the map from S to [0,1] which assigns to every object of S its degree of satisfaction of the features of a_i, the above theorem gives the way in which an F-indistinguishability operator on S -and consequently a pseudo-metric- can be constructed from the applications h_i. Thus, for instance

$$L(a,b)=\inf\{1-|h_i(a)-h_i(b)|\ ;\ 1\leqslant i\leqslant n\}$$

will be a likeness relation on S, as well as

$$P(a,b)=\inf\{\min[1,(h_i(a)\wedge h_i(b))/(h_i(a)\vee h_i(b))]\ ;\ 1\leqslant i\leqslant n\}$$

will be a probabilistic relation. Similar expressions can be obtained for any continuous t-norm.

Now we turn our attention to the use of F-indistinguishability operators in order to define material equivalence in $\underset{\sim}{P}(X)$. To this aim we will assume, as in classical logic, that the indistinguishability that defines this connective is an indistinguishability operator with respect to an ordered semigroup of the form $(\underset{\sim}{P}(X),\underset{\sim}{F},\leqslant)$, where $\underset{\sim}{F}$ is the meet connective induced in $\underset{\sim}{P}(X)$ by the t-norm F, i.e. $\underset{\sim}{F}(\underset{\sim}{A},\underset{\sim}{B})(x)=F(\underset{\sim}{A}(x),\underset{\sim}{B}(x))$ for any x in X. If E is one of such operators, then $E(\underset{\sim}{A},\underset{\sim}{B})$ may be considered as the possibility distribution of being $\underset{\sim}{A}$ and $\underset{\sim}{B}$ indistinguishable, i.e. $E(\underset{\sim}{A},\underset{\sim}{B})(x)$ is the degree in which $\underset{\sim}{A}$ and $\underset{\sim}{B}$ are indistinguishable at the point $x\in X$. Another suggestive interpretation of an indistinguishability operator on a set S with values on $(\underset{\sim}{P}(X),\underset{\sim}{F},\leqslant)$ is obtained when X is assumed to be a set of observers, then for a,b in S, E(a,b) is the possibility distribution of finding a and b indistinguishable by the set of observers X, i.e. E(a,b)(x) is the degree in which the observer x finds a and b indistinguishable. First of all we have that any of such indistinguishability operators is given by a family -indexed by X- of F-indistinguishability operators, that is:

Theorem 3.4. Given a non-empty set S, a map E from SxS into $\underset{\sim}{P}(X)$ is an indistinguishability operator on S at level $\underset{\sim}{A} \in \underset{\sim}{P}(X)$ with respect to $(\underset{\sim}{P}(X), \underset{\sim}{F}, \leqslant)$ if, and only if, $E_x(a,b)=E(\tilde{a},\tilde{b})(x)$ is an F-indistinguishability operator on S at level $\underset{\sim}{A}(x)$ for any $x \in X$.

We remark that this theorem holds if we suppose that the t-norm F also depends on x instead of supposing that the t-norm F is the same for any E_x. In this case, however, we will obtain an indistinguishability operator with respect to the semigroup $(\underset{\sim}{P}(X), \underset{\sim}{\cap}, \leqslant)$, where $\underset{\sim}{\cap}$ is defined by means of $(\underset{\sim}{A} \underset{\sim}{\cap} \underset{\sim}{B})(x)=F_x(\underset{\sim}{A}(x), \underset{\sim}{B}(x))$.

As a corollary of theorem 3.4 and theorem 2.4 we get the following result, where n is a strong negation and F_n^* is the t-conorm dual of F via n, i.e. $F_n^* = n \circ F \circ (n \times n)$:

Theorem 3.5. Given a non-empty set S, a map E from SxS into $\underset{\sim}{P}(X)$ is an indistinguishability operator on S at level $\underset{\sim}{1}$ if, and only if, for any $x \in X$, $m_x(a,b)=n(E(a,b)(x))$ is a generalized pseudo-metric on S with respect to $([0,1], F_n^*, \leqslant)$.

The question is, now, if there exists a particular kind of F-indistinguishability operators which should be used in order to define the material equivalence in $\underset{\sim}{P}(X)$. An answer to this question can be obtained by recalling that the material equivalence in classical and Łukasiewicz logics are defined through their respective standard implication connective by means of $a \leftrightarrow b=(a \rightarrow b) \wedge (b \rightarrow a)$. Thus, if we consider in $\underset{\sim}{P}(X)$ such connective functionnally expressible (i.e. $(\underset{\sim}{A} \rightarrow \underset{\sim}{B})(x) = I(\underset{\sim}{A}(x), \underset{\sim}{B}(x))$, where I is a function from $[0,1] \times [0,1]$ into $[0,1]$, called implication function) the natural extension to $\underset{\sim}{P}(X)$ of the above cases would be given by F-indistinguishability operators of the form

$$E_I(x,y)=F(I(x,y), I(y,x)) \qquad (3.1)$$

being F the continuous t-norm defining the meet in $\underset{\sim}{P}(X)$. E_I will be an F-indistinguishability operator if I is an F'-preorder for some t-norm $F' \leq F$, as it follows from theorem 3.1. Consequently we will get such indistinguishability operators just by determining which implication functions are F-preorders for some t-norm F. We analyze this question for the two main kinds of implication functions which are used in the fuzzy literature: R-implications and S-implications (see Valverde,1982), i.e. implication functions which are defined respectively by $I(x,y)=\widehat{F}(x|y)$, being F a continuous t-norm, and $I(x,y)=G(n(x),y)$, being G a continuous t-conorm and n a strong negation function.

First we notice that if an implication function I satisfies the Identity Principle ($I(x,x)=1$, for any x) and is an F-preorder for some F, then the indistinguishability operator E_I defined by (3.1) is independent of the particular t-norm F, that is, we have the

<u>Theorem 3.6.</u> Let I be an implication function satisfying the Identity Principle, if I is an F-preorder for some t-norm F, then

$$E_I(x,y)=I(\mathrm{Max}(x,y),\mathrm{Min}(x,y))$$

is an F'-indistinguishability operator at level 1 for any t-norm F' such that $F' \leq F$.

That is, for the implication functions I which satisfy the Identity Principle, the F-transitivity of I is a sufficient and necessary condition in order to construct indistinguishability operators by means of (3.1). Next step is to analyze which of such implications are F-transitive. We have:

<u>Theorem 3.7.</u> If $I(x,y)$ is an R-implication obtained from the the t-norm F, then I is F-transitive.

Thus, $E_I(x,y)=\widehat{F}(\mathrm{Max}(x,y)|\mathrm{Min}(x,y))$ is an indistinguishability operator (at level 1) with respect to any t-norm F' such that $F' \leq F$. In what concerns the F-transitivity of the S-implications which satisfy the Identity Principle, we have that the Łukasiewicz implications (i.e. implication functions of the form $I(x,y)=f^{(-1)}(f(x)-f(y))$, where f stands for a continuous and strictly decreasing function from $[0,1]$ into $[0,+\infty)$ with $f(1)=0$ and $f^{(-1)}$ is the pseudo-inverse of f) are F-transitive because they are R-implications too; moreover, we have

<u>Theorem 3.8.</u> Let I be an S-implication satisfying the Identity Principle which is not a Łukasiewicz implication, then there exists no t-norm F such that I is F-transitive.

In what concerns to the S-implications which do not satisfy the Identity Principle, we have the

Theorem 3.9. Let I be an S-implication. If I is F-transitive for some t-norm F, then F is nilpotent and $I(x,y) \leq \hat{F}(x|y)$.

(A t-norm F is nilpotent if for any $x \in (0,1)$ there exists y in $(0,1)$ such that $F(x,y)=0$).

Consequently, given a nilpotent t-norm F, the R-implication given by F is the least upper bound of the S-implications which are F-transitive. We end this section by analyzing two examples of S-implications which are F-transitive but do not satisfy the Identity Principle:

Example 3.1. Let F be a nilpotent t-norm and $I(x,y) = \mathrm{Max}(n(x),y)$, then I is F-transitive if ,and only if, $n(x) \leq \hat{F}(x|0)$ for any x. In this case $F(y,n(y))=0$ and consequently

$$F(\mathrm{Max}(n(x),y),\mathrm{Max}(n(y),z))=\mathrm{Max}(F(n(x),y),F(n(x),z),F(y,z)) \leq \mathrm{Max}(n(x),z).$$

Thus, $E_I(x,y)=F(\mathrm{Max}(n(x),y),\mathrm{Max}(n(y),x))$ is an F-indistinguishability operator at level 0.

Example 3.2. Let $I(x,y)=1-x+xy$, then it is easy to compute that I is F-transitive for the t-norm $F_Ł$, giving in this case the indistinguishability $E_I(x,y)=1-x-y+2xy$, which is nothing but the Bočvar indistinguishability operator extended to an indistinguishability operator on $[0,1]$ at level 1/2.

4. AN APPROACH TO SOME KIND OF "FUZZINESS" BY USING INDISTINGUISHABILITY OPERATORS.

This section is devoted to showing that indistinguishability operators on $\underset{\sim}{P}(X)$ with respect to $(\underset{\sim}{P}(X),\underset{\sim}{F},\leq)$ can also be used to define "fuzziness". Indeed we believe that it is (perhaps) in the line of formulating mathematically in $\underset{\sim}{P}(X)$ other kinds of "equality" different from the pointwise equality (see Trillas,1983) where we have the possibility of founding the roots of fuzziness and its meaning as well as the actual beginning of a new study of vagueness and Booleanity.

After the foundational paper (DeLuca and Termini,1972) a lot of papers on measures of fuzziness were produced (see references in Trillas and Riera,1983) specially about the use of entropies and on the anlysis of their different types, but essentially no new ideas appeared in the first eight years. Yager (1979) introduced the idea that fuzziness comes from a "lack of distinction" between a fuzzy set and its complement

$\overline{A}$=1-A in the setting of Fuzzy Set Theory. In that approach fuzziness is measured by some distances whose election is not clearly related with such a lack of distinction, and although the idea seems very interesting it presuposes the existence of some crispness between A and $\overline{A}$, as if all the region "neither not nor yes" were a broad border separating both the "yes" and "not" regions: for example, to think what Yager says in the case of the fuzzy set of "round natural numbers", and specially with "large round numbers", seems very difficult. Yager's paper is a good first step towards a nice idea that needs more work in the line of clarifying the mathematical translation of "distinctness" and its dependence on the distribution of the values of the membership function: it is not evident that only one family of distances are predestinate to measure fuzziness.

Now we will analyze this question by assuming that the "lack of distinction" above mentioned is given by an indistinguishability operator. For the sake of simplicity we restrict our attention to the case where F is the t-norm $F_Ł$, that is $\underset{\sim}{E}(\underset{\sim}{A},\underset{\sim}{B})=\mathrm{Max}(\underset{\sim}{0},\underset{\sim}{A}+\underset{\sim}{B}-\underset{\sim}{1})$. We first remark that, in such case, by applying theorem 2.4 with the strong negation n(x)=1-x, we get the following:

<u>Theorem 4.1.</u> A map E from $\underset{\sim}{P}(X)\times\underset{\sim}{P}(X)$ into $\underset{\sim}{P}(X)$ is an indistinguishability operator with respect to $(\underset{\sim}{P}(X),\underset{\sim}{F}_Ł,\leqslant)$ if, and only if, $d_x(\underset{\sim}{A},\underset{\sim}{B})=1-E(\underset{\sim}{A},\underset{\sim}{B})(x)$ is a normalized pseudo-metric on $\underset{\sim}{P}(X)$ for any x in X.

That is, any family of pseudo-distances on $\underset{\sim}{P}(X)$ may be used to define indistinguishability in $\underset{\sim}{P}(X)$. Thus, for instance we may use any of the following operators where, again for simplicity, we have considered $d_x=d_{x'}$ for any $x,x'\in X$.

Example 4.1. If $d_x(\underset{\sim}{A},\underset{\sim}{B})=|\underset{\sim}{A}(x)-\underset{\sim}{B}(x)|$, we have $E(\underset{\sim}{A},\underset{\sim}{B})=1-|\underset{\sim}{A}(x)-\underset{\sim}{B}(x)|$, i. e. the equivalence of Łukasiewicz.

Example 4.2. If $d_x(\underset{\sim}{A},\underset{\sim}{B})=|\underset{\sim}{A}(x)-\underset{\sim}{B}(x)|/(1+|\underset{\sim}{A}(x)-\underset{\sim}{B}(x)|)$, the Minkowski metrics, we have $E(\underset{\sim}{A},\underset{\sim}{B})=1/(1+|\underset{\sim}{A}(x)-\underset{\sim}{B}(x)|)$.

Example 4.3. Consider binary expansions $\underset{\sim}{A}(x)=\sum_{i\geqslant 1}\underset{\sim}{A}(x)_i/2^i$, $\underset{\sim}{A}(x)_i\in\{0,1\}$. Then distances $d_x(\underset{\sim}{A},\underset{\sim}{B})=\sum_{i\geqslant 1}(|\underset{\sim}{A}(x)_i-\underset{\sim}{B}(x)_i|/2^i)$ are different from the Euclidean and gives

$$E(\underset{\sim}{A},\underset{\sim}{B})=1-\sum_{i\geqslant 1}(|\underset{\sim}{A}(x)_i-\underset{\sim}{B}(x)_i|/2^i)$$

as associated indistinguishability. We call it the "binary indistinguishability".

Example 4.4. If $d(\underset{\sim}{A},\underset{\sim}{B})= \sup_X |\underset{\sim}{A}(x)-\underset{\sim}{B}(x)|$, is $E(\underset{\sim}{A},\underset{\sim}{B}) =\inf_X(1-|\underset{\sim}{A}(x)-\underset{\sim}{B}(x)|)$, the uniform indistinguishability.

Example 4.5. If X is finite, $X=\{x_1,x_2, \ldots,x_n\}$ and $d(\underset{\sim}{A},\underset{\sim}{B})$ $\{ \sum_{i=1}^{n} [\underset{\sim}{A}(x_i)-\underset{\sim}{B}(x_i)]^p \}^{1/p}$, $p\geqslant 1$, we have

$$E(\underset{\sim}{A},\underset{\sim}{B})=1-\{ \sum_{i=1}^{n} [\underset{\sim}{A}(x_i)-\underset{\sim}{B}(x)_i]^p\}^{1/p},$$

the case p=1 corresponding to the Hamming distance and the p=2 to the Euclidean distance.

Now, generalizing a concept with the same name introduced by Łukasiewicz, we define the <u>falsum</u> of a fuzzy set as the indistinguishability between a fuzzy set and the empty set : $f(\underset{\sim}{A})=E(\underset{\sim}{A},\underset{\sim}{\phi})$. The fuzzy set $f(\underset{\sim}{A})$ will be used mainly for two purposes. First of all we define the fuzziness of $\underset{\sim}{A}\in\underset{\sim}{P}(X)$ by means of $f(\underset{\sim}{A})$ and, secondly, some of its properties give us arguments to choose a particular family of distances d_x. We start by analyzing this second question.

When E is the Łukasiewicz equivalence we have $f(\underset{\sim}{A})(x) =1-\underset{\sim}{A}(x)=\overline{\underset{\sim}{A}}(x)$, i.e. the falsum coincides with the complement. This is not the situation with Minkowski distances, being $E(\underset{\sim}{A}, \underset{\sim}{\phi})(x)=1/(1+\underset{\sim}{A}(x))$, fuzzy set which does not have the properties usually required for a "complement". With the binary indistinguishability we have $E(\underset{\sim}{A}, \underset{\sim}{\phi})(x)=1-\sum_{i\geqslant 1}(\underset{\sim}{A}(x)_i/2^i)=1-\underset{\sim}{A}(x)$, that is $f(\underset{\sim}{A})=\overline{\underset{\sim}{A}}$.

To have $f(\underset{\sim}{A})=\overline{\underset{\sim}{A}}$ it is necessary (and sufficient) that $d_x(\underset{\sim}{A}, \underset{\sim}{\phi})=\underset{\sim}{A}(x)$. In order to give a characterization of the distances in $\underset{\sim}{P}(X)$ which satisfy that property, let D_x be the family of distances in $\underset{\sim}{P}(X)$ defined by $D_x(\underset{\sim}{A},\underset{\sim}{B})=\mathrm{Min}(1,\underset{\sim}{A}(x)+\underset{\sim}{B}(x))$, if $\underset{\sim}{A}\neq\underset{\sim}{B}$, $D_x(\underset{\sim}{A},\underset{\sim}{A})=0$. We have

<u>Theorem 4.2.</u> Let d_x be a distance satisfying $d_x(\underset{\sim}{A}, \underset{\sim}{\phi})=\underset{\sim}{A}(x)$ for any $\underset{\sim}{A}\in\underset{\sim}{P}(X)$. Then $|\underset{\sim}{A}(x)-\underset{\sim}{B}(x)|\leqslant d_x(\underset{\sim}{A},\underset{\sim}{B})\leqslant D_x(\underset{\sim}{A},\underset{\sim}{B})$.

In those conditions we have immediately the general representation given by

<u>Theorem 4.3.</u> If d_x is a family of distances satisfying $d_x(\underset{\sim}{A},\underset{\sim}{\phi}) =\underset{\sim}{A}(x)$, for any x in X and any couple $\underset{\sim}{A},\underset{\sim}{B}$ in $\underset{\sim}{P}(X)$, there exists

$\lambda_{\underset{\sim}{AB}}(x) \in [0,1]$ such that

$$d_x(\underset{\sim}{A},\underset{\sim}{B}) = \lambda_{\underset{\sim}{AB}}(x)\,|\underset{\sim}{A}(x)-\underset{\sim}{B}(x)| + (1-\lambda_{\underset{\sim}{AB}}(x))D_x(\underset{\sim}{A},\underset{\sim}{B}).$$

Consequently, all indistinguishability operators such that $f(\underset{\sim}{A})=\overline{\underset{\sim}{A}}$ are representable by

$$E(\underset{\sim}{A},\underset{\sim}{B}) = 1-\lambda_{\underset{\sim}{AB}}(x)\,|\underset{\sim}{A}(x)-\underset{\sim}{B}(x)| - (1-\lambda_{\underset{\sim}{AB}}(x))D_x(\underset{\sim}{A},\underset{\sim}{B}),$$

which is a very large family of indistinguishabilities. Those formulas are rarely useful due to the unknown function $\lambda_{\underset{\sim}{AB}}(x)$. Following the ideas of (Yager, 1979) we introduce the

<u>Definition 4.1.</u> Given an indistinguishability operator E in $\underset{\sim}{P}(X)$ at level $\underset{\sim}{1}$ with respect to $(\underset{\sim}{P}(X), \underset{\sim}{E}_{Ł}, \leqslant)$, the fuzzy set $\delta\underset{\sim}{A}=E(\underset{\sim}{A}, f(\underset{\sim}{A}))$ is said to be the <u>diffusum</u> of $\underset{\sim}{A} \in \underset{\sim}{P}(X)$, and it shows the indistinguishability degrees between $\underset{\sim}{A}$ and its <u>falsum.</u> When the operator E is such that $f(\underset{\sim}{A})=\overline{\underset{\sim}{A}}$, we will speak of "fuzziness" instead of "diffusum".

Taking $d_x(\underset{\sim}{A},\underset{\sim}{B}) = \{\sum_{i=1}^{n}[\,|\underset{\sim}{A}(x_i)-\underset{\sim}{B}(x_i)|/n]^p\}^{1/p}$, $p \geqslant 1$, we have just the cases considered by Yager. In what follows we limit ourselves to the case of fuzzines. It is $\delta\underset{\sim}{A}=E(\underset{\sim}{A},\overline{\underset{\sim}{A}})$ or $\delta\underset{\sim}{A}(x) = 1-d_x(\underset{\sim}{A},\overline{\underset{\sim}{A}}) = 1-\lambda_{\underset{\sim}{A}}(x)\,|2\underset{\sim}{A}(x)-1| - (1-\lambda_{\underset{\sim}{A}}(x)).\underset{\sim}{1} = \lambda_{\underset{\sim}{A}}(x)(1-2\,|\underset{\sim}{A}(x)-1/2|)$ if $\underset{\sim}{A} \neq \underset{\sim}{1/2}$, and $\delta\underset{\sim}{1/2}=0$, writing $\lambda_{\underset{\sim}{A}}(x)$ for $\lambda_{\underset{\sim}{A\overline{A}}}(x)$. We call $\lambda_{\underset{\sim}{A}}(x)$ the <u>specificity function of $\underset{\sim}{A}$</u>.

The fuzziness corresponding to Euclidean distances is exactly $1-2\,|A(x)-1/2|$. We will write $\delta_e(\underset{\sim}{A})(x) = 1-2\,|\underset{\sim}{A}(x)-1/2|$ and call that function the <u>Euclidean fuzziness</u> of $\underset{\sim}{A}$.

<u>Theorem 4.4.</u> The fuzziness of $\underset{\sim}{A} \in \underset{\sim}{P}(X)$ is $\delta\underset{\sim}{A}(x) = \lambda_{\underset{\sim}{A}}(x)\,\delta_e\underset{\sim}{A}(x)$, the product of the specifity function and the Euclidian fuzziness. It is $\delta\underset{\sim}{A}(x) = 2\lambda_{\underset{\sim}{A}}(x)\underset{\sim}{A}(x)$, if $\underset{\sim}{A}(x) \leqslant 1/2$, $\delta\underset{\sim}{A}(x) = \lambda_{\underset{\sim}{A}}(x)\overline{\underset{\sim}{A}}(x)$, if $\underset{\sim}{A}(x) > 1/2$.

In any case we have $\delta\underset{\sim}{A} \leqslant \delta_e\underset{\sim}{A}$: the Euclidean fuzziness is the greatest fuzziness under the law $\underset{\sim}{E}_{Ł}$. Both fuzzinesses are coincidental if, and only if, the specificity function equals $\underset{\sim}{1}$. On the other hand, it is $\delta\underset{\sim}{A}(x) = \lambda_{\underset{\sim}{A}}(x)$, for any x in X, if, and only if, $\delta_e\underset{\sim}{A}=1$, that happens only when $\underset{\sim}{A}=\underset{\sim}{1/2}$: the specificity function is the fuzziness only for the fuzzy set $\underset{\sim}{1/2}$

<u>Theorem 4.5.</u> The Euclidian fuzziness δ_e verifies the properties of an entropy in the sense of DeLuca-Termini (but relatively to the structure of $\underset{\sim}{P}(X)$).

The formula $\delta A=\lambda_A(.)\delta_e A$ shows that fuzziness can be a very sophisticated function, because of the specificity function that is a unknown function. The study of δA deserves a more acurate research.

5. CONCLUDING REMARK: THE POINCARÉ-MENGER PROBLEM.

In the empirical sciences we are not confronted with a problem of "equality" between any two observables but with the problem of their "indistinguishability". Poincaré (1946) already noticed that non-mathematical <u>continua</u> are often endowed with an empirical relation $\approx$ of indistinguishability that is characterized by the existence of triples (x,y,z) for which $x\approx y$ and $y\approx z$ but <u>not</u> $x\approx z$; $\approx$ is therefore a relation for which the transitive property fails.

Menger (1951) conjectured we could find a mathematical formulation (he suggested a probabilistic one) were transitivity would still hold and so things would come back again into formalizable frame.

Naturally, if the $\approx$ relation can be correctly translated into a classical relation as defined by "$x\approx y$ iff $R(x,y)=1$" and "$x\not\approx y$ iff $R(x,y)=0$", then the only tool a mathematician is able to offer is the <u>transitive closure</u> of R, where R itself will get included. Things, however, are not always so crisp. Thus, in the approximate synonymy case, for instance, we have distinctly seen that the indistinguishability is linguistic and, substantially, a matter of degree. If we had choosen the operator $\rho(t,t')=1-p(\sigma_t\Delta\sigma_{t'})$ as the indistinguishability operator for linguistic terms then the transitive property would assuredly hold with the Łukasiewicz operation $F_Ł$. If we had taken the operator $\rho^*(t,t')=|\sigma_t\cap\sigma_{t'}|/|\sigma_t\cup\sigma_{t'}|$ instead, a more accurate analysis would have been required to ascertain whether some structure exists in [0,1] that allows to translate (i.e. control) transitivity. We can say that the <u>transitivity control problem</u> for an indistinguishability operator translating synonymy in a given linguistic environment (context) is a solved problem in the first instance and a well-formulated problem in the second. Determination of the most appropiate operator obviously falls within the tasks of empirical research in the field of Linguistics.

Actually, for any pair of objects x and y in a universe X a careful study should produce a linguistic label such as x and y are almost indistinguishable, x and y and certainly indistinguishable, x and y may be indistinguishable, and so forth. Such assignments are not arbitrary but the result of an agreement among specialists confronting well-defined experiences in a well-defined environment with the (conceptual or physical) tools avalaible. Let $\mathcal{E}$ be the function that assigns a linguistic label $\mathcal{E}(x,y)$ (out of the set L of such labels) to each pair (x,y). We call such triple $(X, \mathcal{E}, L)$ an <u>empirical indistinguishability structure</u>.

Given a structure $(X, \mathcal{E}, L)$ we define the <u>Poincaré-Menger problem</u> as that of finding: a) a set L endowed with an operation $*$, an order $\leqslant$ and a distinguished element λ; and b) an indistinguishability operator E in X appropiate to the $(L,*,\leqslant)$ and λ, so that the values $E(x,y) \in L$ adequately represent the $\mathcal{E}(x,y)$ labels from the experimental standpoint and have the capability to control the breaking of the transitive chains $\mathcal{E}(x_1,x_2)$, $\mathcal{E}(x_2,x_3)$, ..., $\mathcal{E}(x_{n-1},x_n)$, by means of the $E(x_1,x_2)$, $E(x_2,x_3)$, ..., $E(x_{n-1},x_n)$ values.

The Poincaré-Menger problem for congruence of geometric figures is completely solved by the two-valued operator $E(F,G) = 1-\inf_{\Sigma}\sup_{X} d_o(\mu_{\sigma(F)}(x),\mu_G(x))$. The Poincaré-Menger problem for propositional equivalence in classical two-valued logic is completely solved by the biconditional operator. The Poincaré-Menger problem for propositional equivalence in multiple-valued logics seems well-formulated though some relevant aspects need further clarification as e.g. whether an x satisfying $\lambda \leqslant E(x,y)$ with some y can replace y "salva veritate" in any logical formula and, in particular, whether in Łukasiewicz's $\aleph_1$ logic the problem is solved by letting $1-E(x,y)$ be a normalized pseudodistance (and conversely). The same type of multiple-valued logic, but translated to $[0,1]^X$, seems adequate to attack the Poincaré-Menger problem in clustering applications (Ruspini, 1982).

The latter mention of clustering classification processes reminds again Poincaré (1946b). In the article entitled "What a classification should be?", Poincaré notices that the logic arising in a <u>predicative classification</u> (that is unaltered by the introduction of new objects to be classified) is predictably different than the logic arising in a <u>non-predicative classification</u> (that has to be altered at each new occurrence). Generalized relations given by indistinguishability operators are in the second category, modeling what we could call "classification processes", one of whose representatives is Zadeh's "similarity" model.

ACKNWLEDGEMENT: Authors are very grateful to Dr. T. Sales (Barcelona) for his valuable aid and deep critique.

REFERENCES

Barra, J.R.: 1971, Notions fondamentales de Statistique Mathématique, Dunod, Paris.

DeLuca, A. and S. Termini: 1972, 'A definition of a non-probabilistic entropy in the setting of Fuzzy Set Theory', Inf. and Control, 20(4), 301-312.

Menger, K.: 1951, 'Probabilistic Theory of Relations', Proc. Nat. Acad. Sci. USA, 37, 178-180.

Munitz, M.K. (Ed.): 1971, 'Identity and Individuation', New York Univ. Press.

Neveu, J.: 1964, Bases Mathématiques du Calcul des Probabilités, Masson, Paris.

Ovchinnikov, S.V.: 1982, 'On Fuzzy Relational Systems', Proc. 2nd. World Conf. on Math. at Service of Man, Univ. de las Palmas, Gran Canaria, Spain, 566-568.

Parikh, R.: 1983, 'The Problem of Vague Predicates', in R.S. Cohen and M.W. Wartofsky (Eds.), Language, Logic and Method, Reidel, 241-261.

Poincaré, H.: 1946, El valor de la Ciencia, Espasa-Calpe, Buenos Aires.

Poincaré, H.: 1946b, Ultimos pensamientos, Espasa-Calpe, B. Aires.

Reichenbach, H.: 1947, Elements of Symbolic Logic, MacMillan Co.

Ruspini, E.: 1982, 'Recent Developments in Fuzzy Clustering', in R.R. Yager (Ed.), Fuzzy Set and Possibility Theory: Recent Developments, Pergamon Press, 133-147.

Trillas, E.: 1982, 'Assaig sobre les relacions de indistingibilitat', Actes 1er. Congrés Català Lògica Matemàtica, Barcelona.

Trillas, E.: 1983, 'Sobre la igualdad de conjuntos borrosos', Rev. de la Real Academia de Ciencias, LXXVI/4, 895-899, Madrid.

Trillas, E. and C. Alsina: Introducción a los espacios métricos generalizados, Fund. J. March, Serie Universitaria, Madrid.

Trillas, E. and T. Riera: 1982, 'From Measures of Fuziness to Booleanity Control', in M.M. Gupta and E. Sanchez (Ed.), Fuzzy Information and Decission Processes, North Holland, 3-16.

Ullmo, J.: 1969, La pensée scientifique moderne, Flammarion, Paris.

Valverde, L.: 1982, 'Concerning Implication in Fuzzy Logic', Proc. 2nd. World Conf. Math. at Service of Man, (Supplement) Univ. Las Palmas, Gran Canaria, Spain.

Valverde, L.: 1982b, Contribució a l'estudi dels models matemàtics per a lògiques multivalents, Ph. D. Thesis, Univ. Politècnica de Barcelona.

Zadeh, L.A.: 1971, 'Quantitative Fuzzy Semantics', Inf Sci., 3, 159-176.

Zadeh, L.A.: 1971b,'Similarity Relations and Fuzzy Orderings' Inf. Sci.,3, 177-200.

Dept. Matemàtiques i Estadística.
E.T.S. Arquitectura.
Univ. Politècnica de Barcelona.
Avda. Diagonal, 649
BARCELONA-28.
Spain.

L.A.Zadeh

A THEORY OF COMMONSENSE KNOWLEDGE

ABSTRACT. The theory outlined in this paper is based on the idea that what is commonly called *commonsense knowledge* may be viewed as a collection of *dispositions*, that is, propositions with implied fuzzy quantifiers. Typical examples of dispositions are: *Icy roads are slippery. Tall men are not very agile. Overeating causes obesity. Bob loves women. What is rare is expensive*, etc. It is understood that, upon restoration of fuzzy quantifiers, a disposition is converted into a proposition with explicit fuzzy quantifiers, e.g., *Tall men are not very agile* → *Most tall men are not very agile.*

Since traditional logical systems provide no methods for representing the meaning of propositions containing fuzzy quantifiers, such systems are unsuitable for dealing with commonsense knowledge. It is suggested in this paper that an appropriate computational framework for dealing with commonsense knowledge is provided by *fuzzy logic*, which, as its name implies, is the logic underlying fuzzy (or approximate) reasoning. Such a framework, with an emphasis on the representation of dispositions, is outlined and illustrated with examples.

1. INTRODUCTION

It is widely agreed at this juncture that one of the important problem-areas in AI relates to the representation of commonsense knowledge. In general, such knowledge may be regarded as a collection of propositions exemplified by: *Snow is white. Icy roads are slippery. Most Frenchmen are not very tall. Virginia is very intelligent. If a car which is offered for sale is cheap and old then it probably is not in good shape. Heavy smoking causes lung cancer*, etc. Representation of propositions of this type plays a particularly important role in the design of expert systems [3].

The conventional knowledge representation techniques based on the use of predicate calculus and related methods are not well-suited for the representation of commonsense

H. J. Skala, S. Termini, and E. Trillas (eds.), Aspects of Vagueness, 257–295.

knowledge because the predicates in propositions which represent commonsense knowledge do not, in general, have crisp denotations. For example, the proposition *Most Frenchmen are not very tall* cannot be represented as a well-formed formula in predicate calculus because the sets which constitute the denotations of the predicate *tall* and the quantifier *most* in their respective universes of discourse are fuzzy rather than crisp.

More generally, the inapplicability of predicate calculus and related logical systems to the representation of commonsense knowledge reflects the fact that such systems make no provision for dealing with uncertainty. Thus, in predicate logic, for example, a proposition is either true or false and no gradations of truth or membership are allowed. By contrast, in the case of commonsense knowledge, a typical proposition contains a multiplicity of sources of uncertainty. For example, in the case of the proposition *If a car which is offered for sale is cheap and much more than ten years old then it probably is not in good shape*, there are five sources of uncertainty: (i) the temporal uncertainty associated with the fuzzy predicate *much more than ten years old;* (ii) the uncertainty associated with the fuzzy predicate *cheap;* (iii) the uncertainty associated with the fuzzy predicate *not in good shape;* (iv) the probabilistic uncertainty associated with the event *The car is not in good shape;* and (v) the uncertainty associated with the fuzzy characterization of the probability of the event in question *as probable*.

The approach to the representation of commonsense knowledge which is described in this paper is based on the idea that propositions which characterize commonsense knowledge are, for the most part, *dispositions*, that is, propositions with implied fuzzy quantifiers. In this sense, the proposition *Tall men are not very agile* is a disposition which upon restoration is converted into the proposition *Most tall men are not very agile*. In this proposition, *most* is an explicit fuzzy quantifier which provides an approximate characterization of the proportion of *men who are not very agile* among *men who are tall* [63].

To deal with dispositions in a systematic fashion, we shall employ *fuzzy logic* - which is the logic underlying *approximate* or *fuzzy reasoning* [5, 19, 26, 55, 58]. Basically, fuzzy logic has two principal components. The first component is, in effect, a translation system for representing the meaning of propositions and other types of semantic entities. We shall employ the suggestive term *test-score semantics* to refer to this translation system because it involves an aggregation of the test scores of elastic constraints which are induced by the semantic entity whose meaning is represented.

The second component is an inferential system for arriving at an answer to a question which relates to the information which is resident in a knowledge base. In the present paper, the focus of our attention will be the problem of meaning representation in the context of commonsense knowledge, and our discussion of the inferential component will be limited in scope.[1]

2. MEANING REPRESENTATION IN TEST-SCORE SEMANTICS

Test score semantics is concerned with representation of the meaning of various types of semantic entities, e.g., propositions, predicates, commands, questions, modifiers, etc. However, knowledge, whether commonsense or not, may be viewed as a collection of propositions. For this reason, we shall restrict our discussion of test-score semantics to the representation of the meaning of propositions.[2]

In test-score semantics, as in PRUF [57], a proposition is regarded as a collection of elastic, or, equivalently, fuzzy constraints. For example, the proposition *Pat is tall* represents an elastic constraint on the height of Pat. Similarly, the proposition *Charlotte is blonde* represents an elastic constraint on the color of Charlotte's hair. And, the proposition *Most tall men are not very agile* represents an elastic constraint on the proportion of men who are not very agile among tall men.

In more concrete terms, representing the meaning of a proposition, p, through the use of test-score semantics involves the following steps.

1. Identification of the variables $X_1,\ldots,X_n$ whose values are constrained by the proposition. Usually, these variables are implicit rather than explizit in p.
2. Identification of the constraints $C_1,\ldots,C_m$ which are induced by p.
3. Characterization of each constraint, C_i, by describing a testing procedure which associates with C_i a test score τ_i representing the degree to which C_i is

satisfied. Usually τ_i is expressed as a number in the interval [0,1]. More generally, however, a test score may be a probability/possibility distribution over the unit interval.

4. Aggregation of the partial test scores $\tau_1,\ldots,\tau_m$ into a smaller number of test scores $\overline{\tau}_1,\ldots,\overline{\tau}_k$, which are represented as an *overall vector test score* $\tau = (\overline{\tau}_1,\ldots,\overline{\tau}_k)$. In most cases $k = 1$, so that the overall test scores is a scalar. We shall assume that this is the case unless an explicit statement to the contrary is made.

It is important to note that, in test-score semantics, the meaning of p is represented not by the overall test score τ but the procedure which leads to it. Viewed in this perspective, test-score semantics may be regarded as a generalization of truth-conditional, possible-world and model-theoretic semantics [8, 9, 28, 32]. However, by providing a computational framework for dealing with uncertainty - which the conventional semantical systems disregard - test-score semantics achieves a much higher level of expressive power and thus provides a basis for representing the meaning of a much wider variety of propositions in a natural language.

In test-score semantics, the testing of the constraints induced by p is performed on a collection of fuzzy relations which constitute an *explanatory data-base*, or *ED* for short. A basic assumption which is made about the explanatory database is that it is comprised of relations whose meaning is known to the addressee of the meaning-representation process. In an indirect way, then, the testing and aggregation procedures in test-score semantics may be viewed as a description of a process by which the meaning of p is composed from the meanings of the constituent relations in the explanatory database. It is this explanatory role of the relations in *ED* that motivates its description as an *explanatory database*.

As will be seen in the sequel, in describing the testing procedures we need not concern ourselves with the actual entries in the constituent relations. Thus, in general, the description of a test involves only the frames[3] of the constituent relations, that is, their names, their variables (or attributes) and the domain of each variable. When this is the case, the explanatory database will be referred to as the *explanatory database frame*, or *EDF* for short.

As a simple illustration, consider the proposition

$$p = \textit{Debbie is a few years older than Dana.} \qquad (2.1)$$

In this case, a suitable explanatory database frame may be represented as

$$EDF \triangleq POPULATION\ [Name;\ Age] + FEW\ [Number;\ \mu].$$

which signifies that the explanatory database frame consists of two relations: (a) a nonfuzzy relation *POPULATION* [*Name*; *Age*],which lists names of individuals and their age; and (b) a fuzzy relation *FEW* [*Number*; μ], which associates with each value of Number the degree, μ, to which *Number* is compatible with the intended meaning of *few*. In general, the domain of each variable in the *EDF* is implicitly determined by p, and is not spelled-out explicitly unless it is necessary to do so to define the testing procedure.

As another example, consider the disposition[4]

$$p \triangleq \textit{Snow ist white},$$

which is frequently used in the literature to explain the basic ideas underlying truth-conditional semantics.[5]

To construct an *EDF* for this disposition, we first note that what is generally meant by *Snow is white* is *Usually snow ist white*, in which *usually* may be interpreted as a fuzzy quantifier. Consequently, on the assumption that the proposition

$$p' = \textit{Usually snow is white} \qquad (2.2)$$

is a restoration of p, a natural choice for the *EDF* would be

$$EDF \triangleq WHITE\ [Sample;\ \mu] + USUALLY\ [Proportion;\ \mu]. \qquad (2.3)$$

In this *EDF*, the relation *WHITE* is a listing of samples of snow together with the degree, μ, to which each sample is white, while the relation *USUALLY* defines the degree to which a numerical value of *Proportion* is compatible with the intended meaning of *usually*.

In proposition (2.1), the constrained variable is the difference in the ages of Debbie and Dana. Thus,

$$X \triangleq Age(Debbie) - Age(Dana). \qquad (2.4)$$

The elastic constraint which is induced by p is determined by the fuzzy relation *FEW*. More specifically, let Π_X denote the possibility distribution of X, i.e., the fuzzy set of possible values of X. Then, the constraint on X may be expressed as the *possibility assignment equation*

$$\Pi_X = FEW, \tag{2.5}$$

which assigns the fuzzy relation *FEW* to the possibility distribution of X. This equation implies that

$$\pi_X(u) \triangleq Poss\ \{X=u\} = \mu_{FEW}(u), \tag{2.6}$$

where $Poss\ \{X = u\}$ is the possibility that X may take u as its value; $\mu_{FEW}(u)$ is the grade of membership of u in the fuzzy relation *FEW*; and the function π_X (from the domain of X to the unit interval) is the *possibility distribution function* associated with X.

For the example under consideration, what we have done so far may be summarized by stating that the proposition

$$p \triangleq \textit{Debbie is a few years older than Dana}$$

may be expressed in a canonical form, namely,

$$X \textit{ is } FEW, \tag{2.7}$$

which places in evidence the implicit variable X which is constrained by p. The canonical proposition implies and is implied by the possibility assignment equation (2.5), which defines via (2.6) the possibility distribution of X and thus characterizes the elastic constraint on X which is induced by p.

The foregoing analysis may be viewed as an instantiation of the basic idea underlying PRUF, namely, that any proposition, p, in a natural language may be expressed in the canonical form

$$cf(p) \triangleq X \textit{ is } F, \tag{2.8}$$

where $X = (X_1,\ldots,X_n)$ is an n-ary *focal variable* whose constituent variables $X_1,\ldots,X_n$ range over the universes $U_1,\ldots,U_n$, respectively; F is an n-ary fuzzy relation in the product space $U = U_1 \times \ldots \times U_n$ and $cf(p)$ is an abbreviation for *canonical form of p*. The canonical form, in turn, may be expressed more concretely as the possibility assignment equation

$$\Pi_X = F. \qquad (2.9)$$

which signifies that the possibility distribution of X is given by F. Thus, we may say that p *translates* into the possibility assignment equation (2.9), i.e.

$$p \rightarrow \Pi_X = F \qquad (2.10)$$

in the sense that (2.9) exhibits the implicit variable which is constrained by p and defines the elastic constraint which is induced by p.

When we employ test-score semantics, the meaning of a proposition, p, is represented as a test which associates with each *ED* (i.e., an instantiation of *EDF*) an overall test score τ which may be viewed as the *compatibility* of p with *ED*. This compatibility may be interpreted in two equivalent ways: (a) as the truth of p given *ED*; and (b) as the possibility of *ED* given p. The latter interpretation shows that the representation of the meaning of p as a test is equivalent to representing the meaning of p by a possibility assignment equation.[6]

The connection between the two points of view will become clearer in Section 4, where we shall discuss several examples of propositions representing commonsense knowledge. As a preliminary, we shall present in the following section a brief exposition of some of the basic techniques which will be needed in Section 4 to test the constituent relations in the explanatory database and aggregate the partial test scores.

3. TESTING AND TRANSLATION RULES

A typical relation in *EDF* may be expressed as $R[X_1;\ldots;X_n;\mu]$, where R is the name of the relation;

the $X_i, i = 1,\ldots,n$, are the names of the variables (or, equivalently, the attributes of R), with U_i and u_i representing, respectively, the domain of X_i and its generic value; and μ is the grade of membership of a generic n-tuple $u = (u_1, \ldots, u_n)$ in R.

In the case of nonfuzzy relations, a basic operation on R which is a generalization of the familiar operation of looking up the value of a function for a given value of its argument, is the so-called *mapping operation* [11]. The counterpart of this operation for fuzzy relations is the operation of *transduction* [61].

Transduction may be viewed as a combination of two operations: (a) *particularization*[7], which constrains the values of a subset of variables of R; and *projection*, which reads the induced constraints on another subset of variables of R. The subsets in question may be viewed as the *input* and *output* variables, respectively.

To define particularization, it is helpful to view a fuzzy relation as an elastic constraint on n-tuples in $U_1 \times \ldots \times U_n$, with the μ-value for each row in R representing the degree (or the test score) with which the constraint is satisfied.

For concreteness, assume that the input variables are X_1, X_2 and X_3, and that the constraints on these variables are expressed as canonical propositions. For example

$$X_1 \textit{ is } F$$

and

$$(X_2, X_3) \textit{ is } G,$$

where F and G are fuzzy subsets of U_1, and $U_2 \times U_3$, respectively. Equivalently, the constraints in question may be expressed as

$$\Pi_{X_1} = F$$

and

$$\Pi_{(X_2,X_3)} = G,$$

where Π_{X_1} and $\Pi_{(X_2,X_3)}$ are the respective possibility distributions of X_1 and X_2,X_3. To place in evidence the input constraints, the particularized relation is written as

$$R^* \triangleq R[X_1 \text{ is } F;\ (X_2,X_3) \text{ is } G] \tag{3.1}$$

or, equivalently, as

$$R^* \triangleq R[\Pi_{X_1} = F;\ \Pi_{(X_2,X_3)} = G]. \tag{3.2}$$

As a concrete illustration, assume that R is a relation whose frame is expressed as

$$RICH\ [Name;\ Age;\ Height;\ Weight,\ Sex;\ \mu] \tag{3.3}$$

in which Age, Height, Weight and Sex are attributes of Name, and μ is the degree to which Name is *rich*. In this case, the input constraints might be:

Age is YOUNG
(Height, Weight) is BIG
Sex is MALE

and, correspondingly, the particularized relation reads

$$R^* \triangleq RICH\ [Age \text{ is } YOUNG;\ (Height,\ Weight) \text{ is } BIG;\ Sex \text{ is } MALE]. \tag{3.4}$$

To concretize the meaning of a particularized relation it is necessary to perform a *row test* on each row of R. Specifically, with reference to (3.1), let

$r_t = (u_{1t},\ldots,u_{nt},\mu_t)$ be the tth row of R, where

$u_{1t},\ldots,u_{nt},\mu_t$ are the values of $X_1,\ldots,X_n,\mu$, respectively.

Furthermore, let μ_F and μ_G be the respective membership functions of F and G. Then, for τ_i, the test score for the constraints on X_1 and (X_2,X_3) may be expressed as

$$\tau_{1t} = \mu_F(u_{1t})$$

$$\tau_{2t} = \mu_G(u_{2t},u_{3t}).$$

To aggregate the test scores with μ_t, we employ the min operator $\wedge$[8], which leads to the overall test score for τ_t:

$$\tau_t = \tau_{1t} \wedge \tau_{2t} \wedge \mu_t. \tag{3.5}$$

Then, the particularized relation (3.1) is obtained by replacing each μ_t in τ_t, $t = 1,2,\ldots$, by τ_t. An example illustrating these steps in the computation of a particularized relation may be found in [61].

As was stated earlier, when a fuzzy relation R is particularized by constraining a set of input variables, we may focus our attention on a subset of variables of R which are designated as *output* variables and ask the question: What are the induced constraints on the output variables? As in the case of nonfuzzy relations, the answer is yielded by projecting the particularized relation on the cartesian product of the domains of output variables. Thus, for example, if the input variables are X_2,X_3 *and* X_5, and the output variables are X_1 and X_4, then the induced constraints on X_1 and X_4 are determined by the projection, G, of the particularized relation R^* $[(X_2,X_3,X_5)$ *is* $F]$ on $U_1 \times U_2$. The relation which represents the projection is expressed as in [27][9].

$$G \triangleq {}_{X_1 \times X_2} R\ [(X_2,X_3,X_5) \text{ is } F], \tag{3.6}$$

with the understanding that $X_1 \times X_2$ in (3.6) should be interpreted as $U_1 \times U_2$. In more transparent terms, (3.6) may be restated as the *transduction:*

$$If\ (X_2,X_3,X_5)\ is\ F,\ then\ (X_1,X_2)\ is\ G, \tag{3.7}$$

where G is given by (3.6). Equivalently, (3.7) may be interpreted as the instruction:

$$Read\ (X_1,X_2)\ given\ that\ (X_2,X_3,X_5)\ is\ F. \tag{3.8}$$

For example, the transduction represented by the expression

$$\underset{Name\ \times\ \mu}{RICH}\ [Age\ is\ YOUNG;\ (Height,\ Weight)\ is\ BIG;\ Sex\ is\ MALE]$$

may be interpreted as the fuzzy set of names of rich men who are young and big. It may also be interpreted in an imperative sense as the instruction: Read the name and grade of membership in the fuzzy set of rich men of all those who are young and big.

Remark. When the constraint set which is associated with an input variable, say X_1, is a singleton, say {a}, we write simply

$$X = a$$

instead of X is a. For example,

$$\underset{Name\ \times\ \mu}{RICH}\ [Age = 25;\ Weight = 136,\ Sex = Male]$$

represents the fuzzy set of rich men whose age and weight are equal to 25 and 136, respectively.

Composition of Elastic Constraints

In testing the constituent relations in *EDF*, it is helpful to have a collection of standardized translation rules for computing the test score of a combination of elastic constraints $C_1,\ldots,C_k$ from the knowledge of the test scores of each constraint considered in isolation. For the most part, such rules are *default* rules in the sense that they are intended to be used in the absence of alternative rules supplied by the user.

For purposes of commonsense knowledge representation, the principal rules of this type are the following.[10]

3.1. Rules pertaining to modification

If the test score for an elastic constraint C in a specified context is τ, then in the same context the test score for

$$\text{(a) } \textit{not } C \textit{ is } 1 - \tau \text{ (negation)} \tag{3.9}$$

$$\text{(b) } \textit{very } C \textit{ is } \tau^2 \text{ (concentration)} \tag{3.10}$$

$$\text{(c) } \textit{more or less } C \textit{ is } \tau^{\frac{1}{2}} \text{ (diffusion).} \tag{3.11}$$

3.2. Rules pertaining to composition

If the test scores for elastic constraints C_1 and C_2 in a specified context are τ_1 and τ_2 respectively, then in the same context the test score for

$$\text{(a) } C_1 \textit{ and } C_2 \textit{ is } \tau_1 \wedge \tau_2 \text{ (conjunction), where } \wedge \triangleq \min. \tag{3.12}$$

$$\text{(b) } C_1 \textit{ or } C_2 \textit{ is } \tau_1 \vee \tau_2 \text{ (disjunction), where } \vee \triangleq \max. \tag{3.13}$$

$$\text{(c) } \textit{If } C_1 \textit{ then } C_2 \textit{ is } 1 \wedge (1-\tau_1+\tau_2) \text{ (implication).} \tag{3.14}$$

3.3. Rules pertaining to quantification

The rules in question apply to propositions of the general form *Q A's and B's* where *Q* is a fuzzy quantifier, e.g., most, many, several, few, etc. and *A* and *B* are fuzzy sets, e.g., tall men, intelligent men, etc. As was stated earlier, when the fuzzy quantifiers in a proposition are implied rather than explicit, their suppression may be placed in evidence by referring to the proposition as a *disposition*. In this sense, the proposition *Overeating causes obesity* is a disposition which results from the suppression of the fuzzy quantifier *Most* in the proposition *Most of those who overeat are obese*.

To make the concept of a fuzzy quantifier meaningful, it is necessary to define a way of counting the number of elements in a fuzzy set or, equivalently, to determine its cardinality.

There are several ways in which this can be done [61]. For our purposes, it will suffice to employ the concept of a *sigma-count*, which is defined as follows.

Let F be a fuzzy subset of $U = \{u_1,...,u_n\}$

expressed symbolically as

$$F = \mu_1/u_1+...+\mu_n/u_n = \Sigma_i\mu_i/u_i \tag{3.15}$$

or, more simply, as

$$F = \mu_1\ u_1+...+\mu_n u_n, \tag{3.16}$$

in which the term μ_i/u_i, $i = 1,...,n$, signifies that μ_i is the grade of membership of u_i in F, and the plus sign represents the union.[11]

The sigma-count of F is defined as the arithmetic sum of the μ_i, i.e.

$$Count\ (F) \triangleq \Sigma_i\mu_i,\ i = 1,...,n, \tag{3.17}$$

with the understanding that the sum may be rounded, if need be, to the nearest integer. Furthermore, one may stipulate that the terms whose grade of membership falls below a specified threshold be excluded from the summation. The purpose of such an exclusion is to avoid a situation in which a large number of terms with low grades of membership become count-equivalent to a small number of terms with high membership.

The *relative sigma-count*, denoted by $\Sigma Count(F/G)$, may be interpreted as the proportion of elements of F which are in G. More explicitly,

$$\Sigma Count\ (F/G) = \frac{\Sigma Count\ (F \cap G)}{\Sigma Count\ (G)} \tag{3.18}$$

where $F \cap G$, the intersection of F and G, is defined by

$$F \cap G = \Sigma_i(\mu_B(u_i) \wedge \mu_G(u_i))/u_i,\ i = 1,...,n. \tag{3.19}$$

Thus, in terms of the membership functions of F and G, the relative sigma-count of F in G is given by

$$\Sigma Count\ (F/G) = \frac{\Sigma_i \mu_F(u_i) \wedge \mu_G(u_i)}{\Sigma_i \mu_G(u_i)} \tag{3.20}$$

The concept of a relative sigma-count provides a basis for interpreting the meaning of propositions of the form *Q A's are B's*, e.g., *Most young men are healthy*. More specifically, if the focal variable (i.e., the constrained variable) in the proposition in question is taken to be the proportion of *B's* in *A's*, then the corresponding translation rule may be expressed as

$$Q\ A\text{'s are } B\text{'s} \rightarrow \Sigma Count\ (B/A) \text{ is } Q \tag{3.21}$$

or, equivalently, as

$$Q\ A\text{'s are } B\text{'s} \rightarrow \Pi_X = Q \tag{3.22}$$

where

$$X = \frac{\Sigma_i \mu_A(u_i) \wedge \mu_B(u_i)}{\Sigma_i \mu_A(u_i)} . \tag{3.23}$$

As will be seen in the following section, the quantification rule (3.21) together with the other rules described in this section provide a basic conceptual framework for the representation of commonsense knowledge. We shall illustrate the representation process through the medium of several examples in which the meaning of a disposition is represented as a test on a collection of fuzzy relations in an explanatory database.

4. REPRESENTATION OF DISPOSITIONS

To clarify the difference between the conventional approaches to meaning representation and that described in the present paper, shall consider as our first example the disposition

$$d \triangleq \textit{Snow is white} , \tag{4.1}$$

which, as was stated earlier, is frequently employed as an

illustration in introductory expositions of truth-conditional semantics (see footnote 5).

The first step in the representation process involves a restoration of the suppressed quantifiers in d. We shall assume that the intended meaning of d is conveyed by the proposition

$$p \triangleq \textit{Usually snow is white}, \tag{4.2}$$

and, as an *EDF* for (4.2), we shall use (2.3), i.e.

$$EDF \triangleq WHITE\,[Sample;\mu] + USUALLY\,[Proportion;\mu]. \tag{4.3}$$

Let $S_1,\ldots,S_m$ denote samples of snow and let τ_i, $i = 1,\ldots,m$, denote the degree to which the color of S_i matches white. Thus, τ_i may be interpreted as the test score for the constraint on the color of S_i which is induced by *WHITE*.

Using this notation, the steps in the testing procedure may be described as follows:

1. Find the proportion of samples whose color is white:

$$\rho = \frac{\Sigma\text{Count}\ (WHITE)}{m} \tag{4.4}$$

$$= \frac{\tau_1 + \ldots + \tau_m}{m}$$

2. Compute the degree to which ρ satisfies the constraint induced by *USUALLY*:

$$\tau = {}_{\mu}USUALLY\,[Proportion = \rho] \tag{4.5}$$

In (4.5), τ represents the overall test score and the right-hand member signifies that the relation *USUALLY* is particularized by setting Proportion equal to ρ and projecting the resulting relation on μ. The meaning of d, then, is represented by the test procedure which leads to the value of τ.

Equivalently, the meaning of d may be represented as a possibility assignment equation. Specifically, let X denote the focal variable (i.e., the constrained variable) in p. Then we can write

$$d \rightarrow \Pi_X = USUALLY \tag{4.6}$$

where

$$X \triangleq \frac{1}{m} \; \Sigma Count\ (WHITE).$$

Example 2.

To illustrate the use of translation rules relating to modification, we shall consider the disposition

$$d \triangleq \textit{Frenchmen are not very tall.} \tag{4.7}$$

After restoration, the intended meaning of d is assumed to be represented by the proposition

$$p \triangleq \textit{Most Frenchmen are not very tall.} \tag{4.8}$$

To represent the meaning of p, we shall employ an *EDF* whose constituent relations are:

$$\begin{aligned} EDF \triangleq\ & POPULATION\ [Name;\ Height] + \\ & TALL\ [Height;\ \mu] + \\ & MOST\ [Proportion;\ \mu]. \end{aligned} \tag{4.9}$$

The relation *POPULATION* is a tabulation of *Height* as a function of *Name* for a representative group of Frenchmen. In *TALL*, μ is the degree to which a value of *Height* fits the description *tall*; and in *MOST*, μ is the degree to which a numerical value of *Proposition* fits the intended meaning of *most*.

The test procedure which represents the meaning of p involves the following steps:

1. Let $Name_i$ be the name of ith individual in *POPULATION*.

   For each $Name_i$, $i = 1,\ldots,m$, find the height of $Name_i$:

$$Height\ (Name_i) \triangleq {}_{Height}POPULATION\ [Name = Name_i].$$

2. For each $Name_i$, compute the test score for the constraint induced by *TALL:*
$$\tau_i = {}_{\mu}TALL[Height = Height\ (Name_i)].$$

3. Using the translation rules (3.9) and (3.10), compute the test score for the constraint induced by *NOT VERY TALL:*
$$\tau'_i = 1-\tau_i^2.$$

4. Find the relative sigma-count of Frenchmen who are not very tall:
$$p \triangleq \Sigma Count\ (NOT.VERY.TALL/POPULATION)$$
$$= \frac{\Sigma_i \tau'_i}{m}.$$

5. Compute the test score for the constraint induced by *MOST:*
$$\tau = {}_{\mu}MOST\ [Proportion = p]. \tag{4.10}$$

The test score given by (4.10) represents the overall test score for d, and the test procedure which yields τ represents the meaning of d.

Example 3.

Consider the disposition

$$d \triangleq Overeating\ causes\ obesity \tag{4.11}$$

which after restoration is assumed to read[12]

$$p \triangleq Most\ of\ those\ who\ overeat\ are\ obese. \tag{4.12}$$

To represent the meaning of p, we shall employ an *EDF* whose constituent relations are:

$$EDF \triangleq POPULATION\ [Name;\ Overeat;\ Obese] + MOST\ [Proportion;\ \mu]. \tag{4.13}$$

The relation *POPULATION* is a list of names of individuals, with the variables *Overeat* and *Obese* representing, respectively, the degrees to which *Name* overeats and is obese. In *MOST*, μ is the degree to which a numerical value of *Proportion* fits the intended meaning of *MOST*.

The test procedure which represents the meaning of d involves the following steps.

1. Let $Name_i$ be the name of ith individual in *POPULATION*.

   For each $Name_i$, $i = 1,\ldots,m$, find the degrees to which $Name_i$ overeats and is obese:

$$\alpha_i \triangleq \mu_{OVEREAT}(Name_i) \triangleq {}_{Overeat}POPULATION[Name = Name_i] \tag{4.14}$$

and

$$\beta_i \triangleq \mu_{OBESE}(Name_i) \triangleq {}_{Obese}\ POPULATION\ [Name = Name_i]. \tag{4.15}$$

2. Compute the relative sigma-count of *OBESE* in *OVEREAT*:

$$p \triangleq \Sigma Count(OBESE/OVEREAT) = \frac{\Sigma_i \alpha_i \wedge \beta_i}{\Sigma_i \alpha_i}. \tag{4.16}$$

3. Compute the test score for the constraint induced by *MOST*:

$$\tau = {}_{\mu}MOST\ [Proportion = \rho]. \tag{4.17}$$

This test score represents the compatibility of d with the explanatory database.

Example 4.

Consider the disposition

$$d \triangleq Heavy\ smoking\ causes\ lung\ cancer. \tag{4.18}$$

Although it has the same form as (4.11), we shall interpret it differently. Specifically, the restored proposition will be assumed to be expressed as

$p \triangleq$ *The incidence of cases of lung cancer among heavy smokers is much higher than among those who are not heavy smokers.* (4.19)

The *EDF* for this proposition is assumed to have the following constituents:

$$EDF \triangleq \mathit{POPULATION}\ [\mathit{Name};\ \mathit{Heavy.\ Smoker};\ \mathit{Lung.Cancer}] + \mathit{MUCH.\ HIGHER}\ [\mathit{Proportion\ 1};\ \mathit{Proportion\ 2};\ \mu]. \quad (4.20)$$

In *POPULATION*, *Heavy. Smoker* represents the degree to which *Name* is a heavy smoker, and the variable *Lung. Cancer* is 1 or 0 depending on whether or not *Name* has lung cancer. In *MUCH. HIGHER*, μ is the degree to which *Proportion 1* is much higher than *Proportion 2*.

The steps in the test procedure may be summarized as follows:

1. For each $Name_i$, $i = 1,\ldots,m$, determine the degree to which $Name_i$ is a heavy smoker:

$$\alpha_i \triangleq {}_{Heavy.Smoker}\mathit{POPULATION}\ [Name = Name_i]. \quad (4.21)$$

Then, the degree to which $Name_i$ is not a heavy smoker is

$$\beta_i = 1 - \alpha_i. \quad (4.22)$$

2. For each $Name_i$, determine if $Name_i$ has lung cancer:

$$\lambda_i \triangleq {}_{Lung.Cancer}\mathit{POPULATION}\ [Name = Name_i]. \quad (4.23)$$

3. Compute the relative sigma-counts of those who have lung cancer among (a) heavy smokers; and (b) not heavy smokers:

$$\rho_1 = \Sigma Count\ (\mathit{LUNG.\ CANCER/HEAVY.\ SMOKER})$$

$$= \frac{\Sigma_i \lambda_i \wedge \alpha_i}{\Sigma_i \alpha_i}$$

$$\rho_2 = \Sigma Count\ (LUNG.\ CANCER/NOT.\ HEAVY.\ SMOKER)$$

$$= \frac{\Sigma_i \lambda_i \wedge (1 - \alpha_i)}{\Sigma_i 1 - \alpha_i}$$

4. Test the constraint induced by MUCH. HIGHER:

$$\tau = {}_{\mu}MUCH.HIGHER[Proportion\ 1 = \rho_1; Proportion\ 2 = \rho_2] \quad (4.24)$$

Example 5.

Consider the disposition

$$d \triangleq Small\ families\ are\ friendly \quad (4.25)$$

which we shall interpret as the proposition

$p \triangleq$ *In most small families almost all of the members are friendly with one another.* (4.26)

It should be noted that the quantifier *most* in p is a second order fuzzy quantifier in the sense that it represents a fuzzy count of fuzzy sets (i.e., *small families).*

The *EDF* for p is assumed to be expressed by

EDF $\triangleq$ *POPULATION* [*Name; Family. Identifier*] + (4.27)
SMALL [*Number;* μ] +
FRIENDLY [*Name 1; Name 2;* μ] +
MOST [*Proportion;* μ] +
ALMOST. ALL [*Proportion;* μ] .

The relation *POPULATION* is assumed to be partitioned (by rows) into disjoint families $F_1,\ldots,F_k$. In *FRIENDLY*, μ is the degree to which *Name 1* is friendly toward *Name 2*, *with Name 1* $\neq$ *Name 2*.

The test procedure may be described as follows:

1. For each family, F_i, find the count of its members:

$$C_i \triangleq Count\ POPULATION\ [Family.Identifier = F_i]. \quad (4.28)$$

2. For each family, test the constraint on C_i induced by *SMALL*:

$$\alpha_i \triangleq {}_{\mu}SMALL\ [Number = C_i]. \tag{4.29}$$

3. For each family, compute the relative sigma-count of its members who are friendly with one another:

$$\beta_i = \frac{1}{(C_i^2 - C_i)} \Sigma_{jk}\ ({}_{\mu}FRIENDLY\ \ [Name\ 1 = Name_j; \\ Name\ 2 = Name_k]) \tag{4.30}$$

where $Name_j$ and $Name_k$ range over the members of F_i and $Name_i \neq Name_j$. The normalizing factor $C_i^2 - C_i$ represents the total number of links between pairs of distinct individuals in F_i.

4. For each family, test the constraint on β_i which is induced by *ALMOST. ALL*:

$$\gamma_i = {}_{\mu}ALMOST.\ ALL\ [Proportion = \beta_i]. \tag{4.31}$$

5. For each family, aggregate the test scores α_i and γ_i by using the min operator ($\wedge$):

$$\delta_i \triangleq \alpha_i \wedge \gamma_i. \tag{4.32}$$

6. Compute the relative sigma-count of small families in which almost all members are friendly with one another:

$$\rho = \frac{1}{k}(\delta_1 + \ldots + \delta_k) \tag{4.33}$$

7. Test the constraint on ρ induced by *MOST*:

$$\tau = {}_{\mu}MOST\ [Proportion = \rho]. \tag{4.35}$$

The value of τ given by (4.35) represents the compatibility of d with the explanatory database.

The foregoing examples are intended to illustrate the basic idea underlying our approach to the representation of commonsense knowledge, namely, the conversion of a disposition into a proposition, and the construction of a test procedure which acts on the constituent relations in an explanatory database and yields its compatibility with the restored proposition.

5. INFERENCE FROM DISPOSITIONS[13]

A basic issue which will be addressed only briefly in the present paper is the following. Assuming that we have represented a collection of dispositions in the manner described above, how can an answer to a query be determined from the representations in question? In what follows, we shall consider a few problems of this type which are of relevance to the computation of certainty factors in expert systems [3, 15, 45, 49, 51].

Conjunction of consequents

Consider two dispositions d_1 and d_2 which upon restoration become propositions of the general form

$$d_1 \rightarrow p_1 \triangleq Q_1 \text{ } A\text{'s are } B\text{'s} \tag{5.1}$$

$$d_2 \rightarrow p_2 \triangleq Q_2 \text{ } A\text{'s are } C\text{'s.} \tag{5.2}$$

Now assume that p_1 and p_2 appear as premises in the inference schema[14]

$$\begin{array}{l} Q_1 \text{ } A\text{'s are } B\text{'s} \\ \underline{Q_2 \text{ } A\text{'s are } C\text{'s}} \\ ? \ Q \text{ } A\text{'s are } (B \text{ and } C)\text{'s.} \end{array} \tag{5.3}$$

in which $?Q$ is a fuzzy quantifier which is to be determined. We shall refer to this schema as the *conjunction of consequents*.

As stated in the following assertion, the fuzzy quantifier Q is bounded by two fuzzy numbers. More specifically, on interpreting Q_1 and Q_2 as fuzzy numbers, we can assert that

$$0 \circledvee (Q_1 \oplus Q_2 \ominus 1) \leq Q \leq Q_1 \circledwedge Q_2, \tag{5.4}$$

in which the operators $\circledwedge$, $\circledvee$, $\oplus$, and $\ominus$, and the inequality $\leq$ are the extensions of $\wedge$, $\vee$, $+$, $-$ and $\leq$, respectively, to fuzzy numbers [14].

Proof. We shall consider the upper bound first. To this end, it will suffice to show that

$$\Sigma Count\ (B \cap C/A) \leq \Sigma Count\ (B/A) \wedge \Sigma Count\ (C/A), \tag{5.5}$$

since, in view of (3.21), the fuzzy quantifiers Q, Q_1 *and* Q_2 may be regarded as fuzzy characterizations of the corresponding sigma-counts.

For convenience, let α_i, β_i and δ_i, $i = 1,\ldots,n$, denote, respectively, the grades of membership of μ_i in A, B and C. Then, on using (3.18) - (3.20), we may write

$$\Sigma Count\ (B \cap C/A) = \frac{\Sigma Count((A \cap B) \cap (A \cap C))}{\Sigma Count\ (A)} \tag{5.6}$$

$$= \frac{1}{\Sigma Count(A)} [\Sigma_i (\alpha_i \wedge \beta_i) \wedge (\alpha_i \wedge \delta_i)].$$

Now

$$\Sigma_i (\alpha_i \wedge \beta_i) \wedge (\alpha_i \wedge \delta_i) \leq \Sigma_i (\alpha_i \wedge \beta_i)$$

$$\Sigma_i (\alpha_i \wedge \beta_i) \wedge (\alpha_i \wedge \delta_i) \leq \Sigma_i (\alpha_i \wedge \delta_i)$$

and hence

$$\Sigma_i (\alpha_i \wedge \beta_i) \wedge (\alpha_i \wedge \delta_i) \leq \Sigma_i (\alpha_i \wedge \beta_i) \wedge \Sigma_i (\alpha_i \wedge \delta_i). \tag{5.7}$$

Consequently,

$$\Sigma \mathrm{Count}(B \cap C/A) \leq \frac{1}{\Sigma Count(A)}[\Sigma Count(A \cap B) \wedge \Sigma Count(A \cap C)]$$

$$\leq \Sigma Count(B/A) \wedge \Sigma Count(C/A).$$

which is what we set out to establish.

To deduce the lower bound, we note that for any real numbers a, b, we have

$$a \wedge b = a + b - a \vee b. \tag{5.8}$$

Consequently,

$$\frac{1}{\Sigma Count(A)}[\Sigma_i(\alpha_i \wedge \beta_i) \wedge (\alpha_i \wedge \delta_i)] =$$

$$\frac{1}{\Sigma Count(A)}[\Sigma_i(\alpha_i \wedge \beta_i) + \Sigma_i(\alpha_i \wedge \delta_i) - \Sigma_i((\alpha_i \wedge \beta_i) \vee (\alpha_i \wedge \delta_i))],$$

and since

$$\alpha_i \geq (\alpha_i \wedge \beta_i) \vee (\alpha_i \wedge \delta_i)$$

it follows that

$$\Sigma Count(B \cap C/A) \geq \frac{1}{\Sigma Count(A)}[\Sigma_i \alpha_i \wedge \beta_i + \Sigma_i \alpha_i \wedge \delta_i - \Sigma_i \alpha_i]$$

or, equivalently,

$$\Sigma Count(B \cap C/A) \geq \Sigma Count(B/A) + (C/A) - 1, \tag{5.9}$$

from which (5.4) follows by an application of the extension principle and the observation that the left-hand member of (5.5) must be non-negative [63].

In conclusion, the simple proof given above establishes the validity of the following inference schema, which, for convenience, will be referred to as the *consequent conjunction syllogism:*

$$\begin{array}{l} Q_1 \text{ } A\text{'s are } B\text{'s} \\ \underline{Q_2 \text{ } A\text{'s are } C\text{'s}} \\ Q \text{ } A\text{'s are } (B \text{ and } C)\text{'s.} \end{array} \tag{5.10}$$

where

$$0 \circledvee (Q_1 \oplus Q_2 \ominus 1) \leq Q \leq Q_1 \circledwedge Q_2.$$

As an illustration, from

$$p_1 \triangleq \textit{Most Frenchmen are not tall}$$

$$p_2 \triangleq \textit{Most Frenchmen are not short} \tag{5.11}$$

we can infer that

Q Frenchmen are not tall and not short

where

$$0 \circledvee (2\ most \ominus 1) \leq Q \leq most. \tag{5.12}$$

In the above example, the variable of interest is the proportion of Frenchmen who are *not tall* and *not short*. In a more general setting, the variable of interest may be a specified function of the variables constrained by the knowledge base. The following variation on (5.11) is intended to give an idea of how the value of the variable of interest may be inferred by an application of the extension principle [56].

Example 6.

Infer from the propositions

$$p_1 \triangleq \textit{Most Frenchmen are not tall} \tag{5.13}$$

$$p_2 \triangleq \textit{Most Frenchmen are not short}$$

the answer to the question

$$q \triangleq \textit{What is the average height of a Frenchman?} \tag{5.14}$$

Because of the simplicity of p_1 and p_2, the constraints induced by the premises may be found directly. Specifically, let $h_1,\ldots h_n$ denote the heights of $Frenchman_1,\ldots,Frenchman_n$, respectively. Then, the test scores associated with the constraints in question may be expressed as

$$\tau_1 = \mu_{ANT\ MOST}(\frac{1}{n}\,\Sigma_i \mu_{TALL}(h_i)) \tag{5.15}$$

and

$$\tau_2 = \mu_{ANT\ MOST}(\frac{1}{n}\,\Sigma_i \mu_{SHORT}(h_i)), \tag{5.16}$$

where *ANT* is an abbreviation for *antonym*, i.e.,

$$\mu_{ANT\ MOST}(u) = \mu_{MOST}(1-u), \quad u\varepsilon[0,1], \tag{5.17}$$

and μ_{TALL} and μ_{SHORT} are the membership functions of *TALL* and *SHORT*, respectively. Correspondingly, the overall test score may be expressed as

$$\tau = \tau_1 \wedge \tau_2.$$

Now, the average height of a Frenchman and hence the answer to the question is given by

$$ans(q) = \frac{1}{n}\,\Sigma_i h_i. \tag{5.18}$$

Consequently, the possibility distribution of $ans(q)$ is given by the solution of the nonlinear program

$$\mu_{ans(q)}(h) = \max_{h_1,\ldots,h_n}(\tau) \tag{5.19}$$

subject to the constraint

$$h = \frac{1}{n}\,\Sigma_i h_i. \tag{5.20}$$

Alternatively, a simpler but less informative answer may be formulated by forming the intersection of the possibility distributions of $ans(q)$ which are induced separately by p_1 and p_2. More specifically, let

$\Pi_{ans(q)|p_1}$, $\Pi_{ans(q)|p_2}$, $\Pi_{ans(q)|p_1 \wedge p_2}$ be the possibility distributions of $ans(q)$ which are induced by p_1, p_2, and the conjunction of p_1 and p_2, respectively. Then, by using the minimax inequality [54], it can readily be shown that

$$\Pi_{ans(q)|p_1} \cap \Pi_{ans(q)|p_2} \supset \Pi_{ans(q)|p_1 \wedge p_2} \qquad (5.21)$$

and hence we can invoke the entailment principle [58] to validate the intersection in question as the possibility distribution of $ans(q)$. For the example under consideration, the desired possibility distribution is readily found to be given by

$$Poss\{ans(q)=h\}=\mu_{ANT\ MOST}(\mu_{TALL}(h)) \wedge \mu_{ANT\ MOST}(\mu_{SHORT}(h)). \qquad (5.22)$$

Chaining of dispositions

As in (5.1) and (5.2), consider two dispositions d_1 and d_2, which upon restoration become propositions of the general form

$$d_1 \rightarrow p_1 \triangleq Q_1 \text{ A's are B's}$$

$$d_2 \rightarrow p_2 \triangleq Q_2 \text{ B's are C's.}$$

An ordered pair, (p_1, p_2), of propositions of this form will be said to form a *chain*. More generally, an *n-ary chain* may be represented as an ordered n-tuple

$$(Q_1A_1\text{'s are }B_1\text{'s}, Q_2A_2\text{'s are }B_2\text{'s}, \ldots, Q_nA_n\text{ are }B_n\text{'s}), \qquad (5.23)$$

in which $B_1 = A_2$, $B_2 = A_3, \ldots, B_{n-1} = A_n$.

Now assume that p_1 and p_2 appear as premises in the inference schema

Q_1 *A's are B's*	*(major premise)*	(5.24)
Q_2 *B's are C's*	*(minor premise)*	
?Q A's are C's	*(conclusion)*	

in which *?Q* is a fuzzy quantifier which is to be determined.

A basic rule of inference which is established in [63] and which has a direct bearing - as we shall see presently - on the determination of Q, is the *intersection/product syllogism*

$$\frac{\begin{array}{l} Q_1 \; A\text{'s are } B\text{'s} \\ Q_2 \; (A \text{ and } B)\text{'s are } C\text{'s} \end{array}}{(Q_1 \otimes Q_2 \; A\text{'s are } (B \text{ and } C)\text{'s},} \tag{5.25}$$

in which $Q_1 \otimes Q_2$ is a fuzzy number which is the fuzzy product of the fuzzy numbers Q_1 and Q_2. For example, as a special case of (5.25), we may write

Most students are single (5.26)
<u>*A little more than a half of single students are male*</u>
(Most ⊗ A little more than a half) of students are single and male.

Since the intersection of B and C is contained in C, the following corollary of (5.25) is its immediate consequence

$$\frac{\begin{array}{l} Q \;\; A\text{'s are } B\text{'s} \\ Q_2 \; (A \text{ and } B)\text{'s are } C\text{'s} \end{array}}{\geq (Q_1 \otimes Q_2) \; A\text{'s are } C\text{'s},} \tag{5.27}$$

where the fuzzy number $\geq (Q_1 \otimes Q_2)$ should be read as *at least* $(Q_1 \otimes Q_2)$, with the understanding that $\geq (Q_1 \otimes Q_2)$ represents the composition of the binary nonfuzzy relation $\geq$ with the unary fuzzy relation $(Q_1 \otimes Q_2)$. In particular, if the fuzzy quantifiers Q_1 and Q_2 are monotone nondecreasing (e.g., when $Q_1 = Q_2 \triangleq$ *most*), then as is stated in [63],

$$\geq (Q_1 \otimes Q_2) = Q_1 \otimes Q_2, \tag{5.28}$$

and (5.27) becomes

$$\frac{\begin{array}{l} Q_1 \; A\text{'s are } B\text{'s} \\ Q_2 \; (A \text{ and } B\text{'s are } C\text{'s} \end{array}}{(Q_1 \otimes Q_2) \; A\text{'s are } C\text{'s}} \tag{5.29}$$

There is an important special case in which the premises in (5.29) form a chain. Specifically, if $B \subset A$, then

$$A \cap B = B$$

and (5.29) reduces to what will be referred to as the *product chain rule*, namely,

$$\frac{\begin{array}{l} Q_1 \; A\text{'s are } B\text{'s} \\ Q_2 B\text{'s are } C\text{'s} \end{array}}{(Q_1 \otimes Q_2) \; A\text{'s are } C\text{'s}.} \tag{5.30}$$

In this case, the chain $(Q_1$ A's are B's, Q_2 B's are C's) will be said to be *product transitive*.[15]

As an illustration of (5.30), we can assert that

$$\frac{\begin{array}{l} \textit{Most students are undergraduates} \\ \textit{Most undergraduates are young} \end{array}}{\textit{Most}^2 \textit{ students are young,}}$$

where $Most^2$ represents the product of the fuzzy number *Most* with itself.

Chaining under reversibility

An important chaining rule which is approximate in nature relates to the case where the major premise in the inference chain

$$\frac{\begin{array}{l} Q_1 \; A\text{'s are } B\text{'s} \\ Q_2 \; B\text{'s are } C\text{'s} \end{array}}{Q \; A\text{'s are } C\text{'s}} \tag{5.31}$$

is *reversible* in the sense that

$$Q_1 \ A\text{'s are } B\text{'s} \overset{\sim}{\leftrightarrow} Q_1 \ B\text{'s are } A\text{'s}, \tag{5.32}$$

where $\overset{\sim}{\leftrightarrow}$ denotes approximate semantic equivalence [57]. - For example,

Most American cars are big $\overset{\sim}{\leftrightarrow}$ *Most big cars are American.* (5.33)

Under the assumption of reversibility, the following syllogism holds in an approximate sense

$$Q \ \ A\text{'s are } B\text{'s} \tag{5.34}$$

$$\frac{Q_2 \ B\text{'s are } C\text{'s}}{\geq (0 \vee (Q_1 \oplus Q_2 \ominus 1)) \ A\text{'s are } C\text{'s}.}$$

We shall refer to this syllogism as the *R-rule*.

To demonstrate the approximate validity of this rule we shall first establish the following lemma.

Lemma

$$\text{If } \Sigma Count(A) = \Sigma Count(B) \tag{5.35}$$

and

$$\Sigma Count(B/A) \geq q_1 \tag{5.36}$$

$$\Sigma Count(C/B) \geq q_2 \tag{5.37}$$

then

$$\Sigma Count(C/A) \geq (0 \vee (q_1 + q_2 - 1)). \tag{5.38}$$

Proof.

We have

$$\Sigma Count(B/A) = \frac{\Sigma Count(A \cap B)}{\Sigma Count(A)}$$

and

$$\Sigma Count(C/B) = \frac{\Sigma Count(B \cap C)}{\Sigma Count(B)}$$

$$= \frac{\Sigma Count(B \cap C)}{\Sigma Count(A)}$$

in virtue of (5.35).

For simplicity, we shall denote $\mu_A(u_i)$, $\mu_B(u_i)$ and $\mu_C(u_i)$, $i = 1,\ldots,n$, by α_i, β_i, γ_i, respectively. Then,

$$\frac{\Sigma Count(A \cap B)}{\Sigma Count(A)} = \frac{\Sigma_i\ \alpha_i \wedge \beta_i}{\Sigma_i\ \alpha_i} \geq q_1 \tag{5.39}$$

$$\frac{\Sigma Count(B \cap C)}{\Sigma Count(A)} = \frac{\Sigma_i\ \beta_i \wedge \gamma_i}{\Sigma_i\ \alpha_i} \geq q_2 \tag{5.40}$$

and hence

$$\frac{\Sigma_i(\alpha_i \wedge \beta_i + \beta_i \wedge \gamma_i)}{\Sigma_i\ \alpha_i} \geq q_1 + q_2, \tag{5.41}$$

and

$$\frac{\Sigma_i(\alpha_i \wedge \beta_i + \beta_i \wedge \gamma_i - \alpha_i)}{\Sigma_i\ \alpha_i} \geq q_1 + q_2 - 1. \tag{5.42}$$

Since

$$\frac{\Sigma Count(C \cap A)}{\Sigma Count(A)} = \frac{\Sigma_i\ \alpha_i \wedge \gamma_i}{\Sigma_i\ \alpha_i}, \tag{5.43}$$

it follows from (5.42) and (5.43) that, to establish (5.38), it will suffice to show that

$$\Sigma_i(\alpha_i \wedge \gamma_i) \geq \Sigma_i(\alpha_i \wedge \beta_i + \beta_i \wedge \gamma_i - \alpha_i), \tag{5.44}$$

or, equivalently,

$$\Sigma_i[(\alpha_i - \alpha_i \wedge \beta_i) + (\beta_i - \beta_i \wedge \gamma_i)] \geq \Sigma_i(\alpha_i - \alpha_i \wedge \gamma_i), \tag{5.45}$$

which is a consequence of the equality

$$\Sigma_i\ \alpha_i = \Sigma_i\ \beta_i, \tag{5.46}$$

which in turn follows from (5.35).

Now, to establish (5.45) it is sufficient to show that, for each i, $i = 1,\ldots,n$, we have

$$(\alpha_i - \alpha_i \wedge \beta_i) + (\beta_i - \beta_i \wedge \gamma_i) \geq (\alpha_i - \alpha_i \wedge \gamma_i), \tag{5.48}$$

in which the summands as well as the right-hand member are non-negative. To this end, we shall verify that (5.48) holds for all possible values of α_i, β_i and γ_i.

Case 1.

Consider all values of α_i, β_i and γ_i which satisfy the inequality $\alpha_i \leq \gamma_i$. In this case, the right-hand member of (5.48) is zero and thus the inequality is verified.

Case 2.

$\alpha_i > \gamma_i$. In this case, we shall consider four subcases.

(i) $\alpha_i \leq \beta_i$, $\beta_i \leq \gamma_i$, which contradicts $\alpha_i > \gamma_i$.

(ii) $\alpha_i > \beta_i$, $\beta_i \leq \gamma_i$, which verifies the inequality.

(iii) $\alpha_i \leq \beta_i$, $\beta_i > \gamma_i$, which verifies the inequality.

(iv) $\gamma_i > \beta_i$, $\beta_i > \gamma_i$, which verifies the inequality.

This concludes the proof of the lemma.

Now, if condition (5.35), i.e.,

$$\Sigma_i \alpha_i = \Sigma_i \beta_i,$$

were not needed to prove the lemma, we could invoke the extension principle [56] to extend the inequality

$$\frac{\Sigma Count(C/A)}{\Sigma Count(A)} \geq 0 \vee (q_1 + q_2 - 1),$$

which holds for real numbers, to

$$Q \geq 0 \circledvee (Q_1 \oplus Q_2 \ominus 1), \tag{5.49}$$

which holds for the fuzzy quantifiers in (5.31). As it is, the assumption of reversibility, i.e.,

$$\frac{\Sigma \text{Count}(\text{A} \cap \text{B})}{\Sigma \text{Count}(\text{A})} \text{ is } Q_1$$

$$\frac{\Sigma Count(B \cap A)}{\Sigma Count(B)} \text{ } is \text{ } Q_1$$

implies the equality

$$\Sigma Count(A) = \Sigma Count(B)$$

only in an approximate sense. Consequently, as was stated earlier, the *R-rule* (5.34) also holds only in an approximate sense. The question of how this sense could be defined more precisely presents a nontrivial problem which will not be addressed in this paper.

Concluding Remark

The point of departure in this paper is the idea that commonsense knowledge may be regarded as a collection of dispositions. Based on this idea, the representation of commonsense knowledge may be reduced, in most cases, to the representation of fuzzily-quantified propositions through the use of testscore semantics. Then, the rules of inference of fuzzy logic may be employed to deduce answers to questions which relate to the information resident in a knowledge base.

The computational framework for dealing with commonsense knowledge which is provided by fuzzy logic is of relevance to the management of uncertainty in expert systems. The advantage of employing fuzzy logic in this application-area is that it provides a systematic framework for syllogistic reasoning and thus puts on a firmer basis the derivation of combining functions for uncertain evidence. The consequent conjunction syllogism which we established in Section 5 is, in effect, an example of such a combining function. What it demonstrates, however, is that, in general, combining functions cannot be expected to yield real-valued probabilities or certainty factors, as they do in MYCIN, PROSPECTOR and other expert systems. Thus, in general, the value returned by a combining function should be a fuzzy number or an n-tuple of such numbers.

Computer Science Division, Department of EECS and ERL;
University of California, Berkeley.

ACKNOWLEDGMENT

This paper is dedicated to Professor Hans J. Zimmermann.

NOTES

1. A more detailed discussion of the inferential component of fuzzy logic may be found in [5, 58, 59]. Recent literature, [26], contains a number of papers dealing with fuzzy logic and its applications. Descriptions of implemented fuzzy-logic-based inferential systems may be found in [2], [22] and [36].
2. A more detailed exposition of test-score semantics may be found in [61].
3. In the literature of database management systems, some authors employ the term *schema* in the same sense as we employ *frame*. More commonly, however, the term schema is used in a narrower sense [11], to describe the frame of a relation together with the dependencies between the variables.
4. As was stated earlier, a *disposition* is a proposition with implied fuzzy quantifiers. (Note that this definition, too, is a disposition.) For example, the proposition *Small cars are unsafe* is a disposition, since it may be viewed as an abbreviation of the proposition *Most small cars are unsafe*, in which *most* is a fuzzy quantifier. In general, a disposition may be restored in more than one way.
5. In truth-conditional semantics, the truth condition for the proposition *Snow is white* is described as snow is white, which means, in plain terms, that the proposition *Snow is white* is true if and only if snow is white.
6. As is pointed out in [61], the translation of p into a possibility assignment equation is an instance of a *focused* translation. By contrast, representation of the meaning of p by a test on *EDF* is an instance of an *unfocused* translation. The two are equivalent in principle but differ in detail.
7. In the case of nonfuzzy relation, particularization is usually referred to as *selection* [11], or *restriction*.
8. Here and elsewhere in the paper the aggregation operation min ($\wedge$) is used as a default choice when no alternative (e.g., arithmetic mean, geometric mean, etc.) is specified.
9. If R is a fuzzy relation its projection on $U_1 \times U_2$ is obtained by deleting from R all columns other than X_1 and X_2, and forming the union of the resulting tuples.

10. A more detailed discussion of such rules in the context of PRUF may be found in [57].
11. In most cases, the context is sufficient to resolve the question of whether a plus sign should be interpreted as the union or the arithmetic sum.
12. It should be understood that (4.12) is just one of many possible interpretations of (4.11), with no implication that it constitutes a prescriptive interpretation of causality. See [48] for a thorough discussion of relevant issues.
13. Inference from dispositions may be viewed as an alternative approach to default reasoning and non-monotinic logic [27, 29, 30, 40].
14. This schema has a bearing on the rule of combination of evidence for conjunctive hypotheses in MYCIN [46].
15. More generally, an n-ary chain $(Q_1\ A_1\text{'s are } B_1\text{'s},\ldots,Q_n\ A_n\text{'s are } B_n\text{'s})$ will be said to be *product transitive* if from the premises which constitute the chain it may be inferred that $\geq (Q_1 \otimes \ldots \otimes Q_n)\ A_1$'s are B_n's.

REFERENCES

[1] Adams, E.W., H.F. Levine: 1975, 'On the uncertainties transmitted from premises to conclusions in deductive inferences', Synthese 30, 429-460.

[2] Baldwin, J.F., S.Q. Zhou: 1982, 'A fuzzy relational inference language', Report No. EM/FS 132, University of Bristol.

[3] Barr, A., E.W. Feigenbaum: 1982, 'The Handbook of Artificial Intelligence', vols. 1,2 and 3. Los Altos: Kaufmann.

[4] Barwise, J., R. Cooper: 1981, 'Generalized quantifiers and natural language', Linguistics and Philosophy 4, 159-219.

[5] Bellman, R.E., L.A. Zadeh: 1977, 'Local and fuzzy logics', in: Modern Uses of Multiple-Valued Logic, Epstein, G. (ed.), Dordrecht: Reidel, 103-165.

[6] Blanchard, N.: 1981, 'Theories cardinales et ordinales des ensembles flou: les multiensembles', Thesis, University of Claude Bernard, Lyon.

[7] Brachman, R.J., B.C. Smith: 1980, 'Special Issue on Knowledge Representation, SIGART 70.

[8] Cooper, W.S.: 1978, 'Logical Linguistics', Dordrecht: Reidel.

[9] Cresswell, M.J.: 1973, 'Logic and Languages', London: Methuen.

[10] Cushing, S.: 1982, 'Quantifier Meanings - A Study in the Dimensions of Semantic Competence', Amsterdam: North-Holland.

[11] Date, C.J.: 1977, 'An Introduction to Database Systems', Reading: Addison-Wesley.

[12] Davis, R., D.B. Lenat: 1982, 'Knowledge-Based Systems in Artificial Intelligence', New York: McGraw-Hill.

[13] DeLuca, A., S.A. Termini: 1972, 'A definition of non-probabilistic entropy in the setting of fuzzy sets theory', Information and Control 20, 301-312.

[14] Dubois, D., H. Prade: 1980, 'Fuzzy Sets and Systems: Theory and Applications', New York: Academic Press.

[15] Duda, R.O., P.E. Hart, K. Konolige, R. Reboh: 1979, 'A computer-based consultant for mineral exploration', Tech. Report, SRJ International, Menlo Park, CA.

[16] Feigenbaum, E.A.: 1977, 'The art of artificial intelligence: Themes and case studies in knowledge engineering', Proc. IJCAI5.

[17] Findler, N.V. (ed.): 1979, 'Associative Networks: Representation and Use of Knowledge by Computer', New York: Academic Press.

[18] Gaines, B.R.: 1979, 'Logical foundations for database systems', Int. J. Man-Machine Studies 11, 481-500.

[19] Goguen, J.A.: 1969, 'The logic of inexact concepts', Synthese 19, 325-373.

[20] Hofmann, T.R.: 1982, 'Qualitative terms for quantity', In Proc. 6th European Meeting on Cybernetics and Systems Research, R. Trappl, (ed.), Amsterdam: North-Holland.

[21] Hoover, D.N.: 1978, 'Probability logic', Annals Math. Lgoic 14, 287-313.

[22] Ishizuka, M., K.S. Fu, J.T.P. Yao: 1982, 'A rule-based inference with fuzzy set for structural damage assessment', Fuzzy Information and Decision Processes, M. Gupta, and E. Sanchez (eds.), Amsterdam: North-Holland.

[23] Keenan, E.L.: 1971, 'Quantifier structures in English', Foundations of Language 7, 255-336.

[24] Klement, E.P., W. Schwyhla, R. Lowen: 1981, 'Fuzzy probability measures', Fuzzy Sets and Systems 5, 83-108.

[25] Klement, E.P.: 1981, 'Operations on fuzzy sets and fuzzy numbers related to triangular norms', Proc. 11th Conf. Multiple-Valued Logic, Univ. of Oklahoma, Norman, 218-225.

[26] Mamdani, E.H., B.R. Gaines: 1981, 'Fuzzy Reasoning and its Applications', London: Academic Press.

[27] McCarthy, J.: 1980, 'Circumscription: A non-monotonic inference rule', Artificial Intelligence 13, 27-40.

[28] McCawley, J.D.: 1981, 'Everything that Linguists have Always Wanted to Know about Logic', Chicago: University of Chicago Press.

[29] McDermott, D.V., J. Doyle: 1980, 'Non-monotonic logic I', Artificial Intelligence 13, 41-72.

[30] McDermott, D.V.: 1982, 'Non-monotonic logic, II: non-monotonic modal theories', J. Assoc. Comp. Mach. 29, 33-57.

[31] Michalski, R., J.G. Carbonell, T.M. Mitchell: 1983, 'Machine Learning', Palo Alto: Tioga Press.

[32] Miller, G.A., P.N. Johnson-Laird: 1976, 'Language and Perception', Cambridge: Harvard University Press.

[33] Mizumoto, M., M. Umano, K. Tanaka: 1977, 'Implementation of a fuzzy-set-theoretic data-structure system', 3rd Int. Conf. Very Large Data Bases, Tokyo.

[34] Mizumoto, M., S. Fukame, K. Tanaka: 1978, 'Fuzzy reasoning Methods by Zadeh and Mamdani, and improved methods', Proc. Third Workshop on Fuzzy Reasoning, Queen Mary College, London.

[35] Moisil, G.C.: 1975, 'Lectures on the Logic of Fuzzy Reasoning', Bucarest: Scientific Editions.

[36] Noguchi, K., M. Umano, M. Mizumoto, K. Tanaka: 1976, 'Implementation of fuzzy artificial intelligence language FLOU', Technical Report on Automation and Language of IECE.

[37] Orlov, A.I.: 1980, 'Problems of Optimization and Fuzzy Variables', Moscow: Znaniye.

[38] Partee, B.: 1976, 'Montague Grammar', New York: Academic Press.

[39] Peterson, P.: 1979, 'On the logic of few, many and most', Notre Dame J. Formal Logic 20, 155-179.

[40] Reiter, R., G. Criscuolo: 1983, 'Some representational issues in default reasoning', Computers and Mathematics 9, 15-28.

[41] Rescher, N.: 1976, 'Plausible Reasoning', Amsterdam: Van Gorcum.

[42] Rich, E.: 1983, 'Artificial Intelligence', New York: McGraw-Hill.
[43] Rieger, C.: 1975, 'Conceptual memory', in Conceptual Information Processing, Schank, R. (ed.), Amsterdam: North-Holland.
[44] Schubert, L.K., R.G. Goebel, N. Cercone: 1979, 'The structure and organization of a semantic net for comprehension and inference', in: Associative Networks, Findler, N.V., (ed.), New York: Academic Press, 122-178.
[45] Shortliffe, E.H., B. Buchanan: 1975, 'A model of inexact reasoning in medicine', Mathematical Biosciences 23, 351-379.
[46] Shortliffe, E.H.: 1976, 'Computer-Based Medical Consultations: MYCIN', New York: American Elsevier.
[47] Sugeno, M.: 1977, 'Fuzzy measures and fuzzy integrals: a survey', in Fuzzy Automata and Decision Processes, Gupta, M.M., Saridis, G.N. and Gaines, B.R. (eds.) Amsterdam: North Holland, 89-102.
[48] Suppes, P.: 1970, 'A Probabilistic Theory of Causality', Amsterdam: North-Holland.
[49] Szolovits, P., S.G. Pauker: 1978, 'Categorical and probabilistic reasoning in medical diagnosis', Artificial Intelligence 11, 115-144.
[50] Terano, T., M. Sugeno: 1975, 'Conditional fuzzy measures and their applications', Fuzzy Sets and Their Applications to Cognitive and Decision Processes, Zadeh, L.A., Fu, K.S., Tanaka, K., Shimura, M. (eds.), New York: Academic Press, 151-170.
[51] Weiss, S.M., C.A. Kulikowski: 1979, 'EXPERT: A system for developing consultation models', Proc. IJAI6.
[52] Yager, R.R.: 1980, 'Quantified propositions in a linguistic logic', in: Proceedings of the 2nd International Seminar on Fuzzy Set Theory, Klement, E.P. (ed.), Johannes Kepler University, Linz, Austria.
[53] Zadeh, L.A.: 1968, 'Probability measures of fuzzy events', J. Math. Anal. Appl. 23, 421-427.
[54] Zadeh, L.A.; 1971, 'Similarity relations and fuzzy orderings', Inf. Sci. 3, 177-200.
[55] Zadeh, L.A.: 1975, 'Fuzzy logic and approximate reasoning' (in memory of Grigore Moisil), Synthese 30, 407-428.
[56] Zadeh, L.A.: 1975, 'The concept of a linguistic variable and its application to approximate reasoning', Inform. Sci. 8, 199-249, 301-357, 9, 43-80.

[57] Zadeh, L.A.: 1978, 'PRUF - a meaning representation language for natural languages', Int. J. Man-Machine Studies 10, 395-460.

[58] Zadeh, L.A.: 1979, 'A theory of approximate reasoning' Electronics Research Laboratory Memorandum M77/58 University of California, Berkeley. Also in: Machine Intelligence 9, Hayes, J.E., Michie, D., Kulich, L.I. (eds.), New York: Wiley 149-194.

[59] Zadeh, L.A.: 1980, 'Inference in fuzzy logic', Proc. 10th Inter. Symp. on Multiple-valued Logic, Northwestern University, 124-131.

[60] Zadeh, L.A.: 1980, 'Possibility theory and soft data analysis', Electronic Res. Laboratory Memorandum M79/59 University of California, Berkeley; also in Mathematical Frontiers of the Social and Policy Sciences, Cobb, L., and Thrall, R.M. (eds.), Boulder: Westview Press, 69-129 (1981).

[61] Zadeh, L.A.: 1981, 'Test-score semantics for natural languages and meaning-representation via PRUF', Techn. Note 247, AI Center, SRI International, Menlo Park, CA. Also in Empirical Semantics, Rieger, B.B., (ed.), Bochum: Brockmeyer, 281-349.

[62] Zadeh, L.A.: 1981, 'Fuzzy probabilities and their role in decision analysis', Proc. 4th MIT/ONR Workshop on Command, Control and Communications M.I.T., 159-179.

[63] Zadeh, L.A.: 1983, 'A computational approach to fuzzy quantifiers in natural languages', Computers and Mathematics 9, 149-184.

[64] Zimmer, A.: 1982, 'Some experiments concerning the fuzzy meaning of logical quantifiers', in General Surveys of Systems Methodology, Troncoli, L. (ed.), 435-441. Society for General Systems Research, Louisville.

[65] Zimmermann, H.-J., P. Zysno: 1980, 'Latent connectives in human decision making', Fuzzy Sets and Systems 4, 37-52.

INDEX OF NAMES

INDEX OF SUBJECTS